Reflections on Water

American and Comparative Environmental Policy
Sheldon Kamieniecki and Michael E. Kraft, editors

Russell J. Dalton, Paula Garb, Nicholas P. Lovrich, John C. Pierce, and John M. Whiteley, *Critical Masses: Citizens, Nuclear Weapons Production, and Environmental Destruction in the United States and Russia*

Daniel A. Mazmanian and Michael E. Kraft, editors, *Toward Sustainable Communities: Transition and Transformations in Environmental Policy*

Elizabeth R. DeSombre, *Domestic Sources of International Environmental Policy: Industry, Environmentalists, and U.S. Power*

Kate O'Neill, *Waste Trading among Rich Nations: Building a New Theory of Environmental Regulation*

Joachim Blatter and Helen Ingram, editors, *Reflections on Water: New Approaches to Transboundary Conflicts and Cooperation*

Reflections on Water

New Approaches to Transboundary Conflicts and Cooperation

edited by Joachim Blatter and Helen Ingram

The MIT Press
Cambridge, Massachusetts
London, England

This book was set in Sabon by Asco Typesetters, Hong Kong, in QuarkXPress. Printed and bound in the United States of America.

Library of Congress Cataloging-in-Publication Data

Reflections on water: new approaches to transboundary conflicts and cooperation / edited by Joachim Blatter and Helen Ingram.
    p.   cm.—(American and comparative environmental policy)
  Includes bibliographical references and index.
  ISBN 0-262-02487-X (hc.: alk. paper)—ISBN 0-262-52284-5 (pbk.: alk. paper)
  1. Water-supply—Management—International cooperation—Case studies. 2. Water rights—Case studies. 3. Conflict management—Case studies. 4. Environmental policy—International cooperation—Case studies. I. Blatter, Joachim, 1966– II. Ingram, Helen M., 1937– III. Series.

TD345.R44   2001
333.91′17—dc21   00-056862

Dedicated in memory of Albert E. Utton, an inspiration for cross-border cooperation

Contents

Tables

Figures

Foreword

This is the fifth volume published in our MIT Press book series American and Comparative Environmental Policy. *Reflections on Water* represents a major interdisciplinary and collaborative effort to advance our understanding of one of the most basic requisites for life on the planet, access to fresh water. The book offers eight diverse and intriguing case studies from around the world to illustrate new approaches to understanding water resource issues, particularly in transboundary settings. The editors and contributors focus on the multiple meanings of water and the effects of such perspectives on policymaking through network analysis, discourse analysis, historical and ethnographic analysis, and the lenses of social ecology. In combination, these approaches lead to a rich understanding of the multiplicity of forces affecting conflicts over transboundary water resources. Such knowledge can substantially enhance the capacity of citizens and policymakers to forge public policies grounded in sustainable development, which itself seeks ambitiously to integrate environmental, economic, and social goals and values.

At a time when water scarcity is increasing around the world and approaching crisis conditions in many regions, this book's emphasis on developing integrative and interdisciplinary approaches to studying water resources and conflicts is most valuable. The authors do not seek to replace more conventional approaches to water resources and management but rather to supplement the considerable knowledge we have gained from engineering, economics, and legal analysis. They hope to provoke scholars and decision makers in the international water community to reconceptualize water and its management by giving greater emphasis to social and cultural issues. The presentation of this argument through elaborate case studies illustrates well the potential of this new scholarship to assist citizens and decision makers in dealing with not only water resources but other complex environmental challenges as well. For the skeptics, the book includes both an introduction that places the new approaches and methods within the context of changing global needs and expectations and a final chapter that underscores some of the limitations of these approaches as well as continuing research needs.

Reflections on Water illustrates well the kind of works we will include in the series. We intend to publish manuscripts that examine a broad range of environmental policy issues. We are particularly interested in books that focus on interdisciplinary research as well as on the links between policy and environmental problems, issues, and controversies in either American or cross-national settings. Future volumes will analyze the policy dimensions of relationships between humans and the environment from either an empirical or a theoretical perspective. The series will include works that assess environmental policy successes and failures, evaluate new institutional arrangements and policy approaches, and help clarify new directions for environmental policy. We plan to publish high-quality scholarly studies that are written for a wide audience that includes academics, policymakers, environmental scientists and professionals, business and labor leaders, environmentalists, and students concerned with environmental issues. We hope that these books contribute to people's understanding of the most important environmental problems, issues, and policies that society now faces and with which it must deal well into the twenty-first century.

Sheldon Kamieniecki, University of Southern California
Michael E. Kraft, University of Wisconsin–Green Bay
Series Editors

Preface

All life is touched by water, and humans as diverse as poets and military strategists draw inspiration from it. How unfortunate, therefore, that narrow perspectives and a limited range of disciplines dominate water research. This book aims to liberate water from excessively rational and utilitarian mindsets. Water has always had an emotional and symbolic value for communities, and it increasingly provides impetus for the formation of transnational networks and discourses. We explore here the multiple meanings of water in a variety of transboundary settings in the contemporary global context.

For an edited volume to transcend the frequent failings of uneven contributions and lack of coherent focus, the book must emerge from a common research endeavor. Such efforts take a great deal of time and support. This project has a lengthy history and many contributors and supporters, a number of whom are neither editors nor authors. Although the genesis of the project extended even further back in time, the effort to develop new approaches to transboundary water problems got a firm start in a research conference at the Bellagio Study and Conference Center, June 2–6, 1997. This exceptionally fruitful meeting was orchestrated by the late Albert E. Utton, Director of the International Transboundary Resources Center, to whom this book is dedicated. The conference received funding from the Hewlitt and Ford Foundations as well as the Rockefeller Foundation.

Grants from the University of California Institute on Global Peace and Cooperation and the University of California at Irvine Global Peace and Cooperation Studies funded the majority of the research, travel, and translation. The Focused Research Group on International Environment in the School of Social Ecology at the University of California, Irvine provided a continuing forum for intellectual discussion of the project. The editors are particularly grateful to one member of the focused research group, Richard Perry, whose wise counsel, excellent critiques, and scholarly networks linked us with important ideas and chapter contributors. María Rosa García-Acevedo wrote her chapter during a post-doctoral year funded through the Sense of Place Project by the Ford Foundation.

The book manuscript, once about double its present length, went through multiple rounds of rewriting and editing. Pamela Doughman and Suzanne Levesque helped mightily in manuscript preparation, working closely with Dianne Christianson in word processing. Helen Ingram is particularly indebted to Michael Brewster, who coordinated the final round of edits. The editors and authors are most grateful to Sheldon Kamieniecki and Michael Kraft, the series editors whose insights led to important improvements in the text; to the anonymous reviewers whose criticisms were invaluable; and to Clay Morgan, the Acquisitions Editor in Environmental Sciences at MIT Press.

I

Concepts and Meanings

1

Emerging Approaches to Comprehend Changing Global Contexts

Joachim Blatter, Helen Ingram, and Pamela M. Doughman

Water can enlarge perception and challenge the mind. For instance, consider the way objects on the floor of a pond seen through water appear clear and sharp, their colors bright. At the same time, images reflected from the surface of the pond can mirror perfectly the surrounding environment of sky, clouds, vegetation, and the very eyes of the observer. Yet water is more than a passive lens or looking glass. Water is an active agent, changing all it touches. Water cuts canyons into the surface of the earth, revealing the world's most distant past. Water surges forward, creating new courses and possibilities yet to be appreciated by humans.

In spite of the transformational possibilities of water, water is usually framed as a rather uninteresting issue and consigned to certain fixed, disciplinary frames for analysis. It is not that water is widely thought to be unimportant. On the contrary, water is proclaimed to be the next global crisis in the popular media and professional literature (Postel 1999). Instead, its full potential as a subject for study is not being realized.

The purpose of this book is to unbind water from its present subject matter constraints and to call attention to the ways water research can reveal contemporary challenges to modes of governance and ways of thinking. The focus here is upon *transboundary* water, which includes border crossings of several types beyond those of political jurisdiction that the term usually implies. New kinds of structures and relationships are augmenting control over water through modern nation-states and large bureaucratic structures. As we will argue later in this chapter, contemporary water is properly placed in a world of flows where influence is streaming simultaneously toward global and local levels, while at the same time nations retain significant influence pools. Other equally significant and associated transitions are in progress. The hegemony of modern conceptions of the world (the ontological basis of reasoning) and of gathering information and providing proof (the epistemological basis of scholarship) is being forced to make way. We point out the limitations of instrumental rationality in capturing the meanings of water and the shortcomings of modern science in improving our understanding of its treatment in society. This book is meant to stimulate scholars in the international water management community to think thoroughly about the inadequacies of existing approaches and to entertain the possibilities of the alternatives we explore.

In opening up the ontological basis for analyzing transboundary water policy, we intend to complement rather than to supplant modern approaches. Having stressed this complementary role, we nevertheless realize that our endeavor represents a direct challenge to modern scholarship on a more abstract methodological level. Whereas modernist science insists on testing explanatory approaches against each other, we propose a complementary usage of approaches combining inductive and deductive or hermeneutical and scientific approaches. We insist that the researcher must first understand the meaning of water as it exists in a particular local place or social context. Only then can the scholar apply specific explanatory approaches. This priority given to understanding leads us to propose specific methods that concentrate on the social construction of meanings of water as well as that of identities and preferences of actors and communities.

We do not recommend particular transmodern meanings of water as being superior to modern meanings. Moreover, as we will argue in chapter 2, modern meanings are probably less problematic with respect to forging compromises across various political boundaries than are pre- or postmodern meanings. However, we strongly criticize modern reductionist scholarship that neglects the prominence or even the existence of meanings of water, which are beyond human control and rational calculation. The purpose of chapter 2 is to explore the full range of meanings of water that need to be considered in studying transboundary water issues. Before we can embark on that enterprise, it is first necessary to set out the conceptual and contextual logic upon which we base our argument for exploring emerging approaches.

After briefly surveying the current debate about the divergent transformational trends at the turn of the millennium, we assert that the phrase "glocalization" (Robertson 1995)[1] best encapsulates contemporary trends. Subnational units have become unbounded or "disembedded" from nation-states, spawning a plethora of new political entities, mostly in the form of connections and networks, but occasionally in the form of sovereign states. Thus the global and local levels are becoming more directly interconnected. In consequence, the privileged position of the national level as gatekeeper between the domestic and international political arenas has diminished along with its dominance as a unit of reference for personal identity. We show the ways in which the current transformations in the economic, social, and political spheres are challenging the territorially defined, sovereign nation-state.

We similarly question other cornerstones of the modernity project, such as the belief in human rationality and human control. It is no coincidence that when a challenge is made to the sovereign nation-state, conceived as the hierarchical center responsible for steering and controlling social and economic processes, the underlying ontology based on the assumption of human control is also challenged, lessening its credibility. This leads us to develop a scheme that shows the transformation of ontologies over time. The basic message is that the current time is characterized by the puzzling phenomenon that different ontologies coexist, with each claiming credibility: premodern ontologies, which are based on the assumption of a single "objective world" to which mankind has to adapt; modern ontologies putting the individual human onto center stage; and postmodern ontologies proposing multiple realities constructed by social/cultural processes (see figure 1.1). This means to propose a "second-order relativity" that contains the paradoxical claim that it makes sense to give credibility to approaches that conceive the world as "objective," leading to methods based on a strict "testing" logic formulated by Karl Popper (Popper 1996), and at the same time to accept approaches that claim that we can only "read," "interpret," and "construct" the subjective world(s) of actors or groups of actors.

One important consequence of this conceptual cornerstone of this book is that we propose a widening of the legitimate scientific approaches and methods employed in the study of transboundary water policy. We argue that the modern legal, technical, and economic approaches employed in the study of transboundary water issues should be complemented by approaches from the life sciences and the humanities. As life sciences are already widely accepted and incorporated in the processes of water policy, this book concentrates on approaches that promise to enhance the understanding of cultural influences on transboundary water policy. A further, more fundamental reason to propose mainly qualitative in-depth studies has already been mentioned: An understanding of the meaning(s) of water held by the involved actors has to be the first step in any investigation of cross-border water politics. Otherwise, perceptions of and connections to water, which are not useful in the modern sense, are systematically neglected. We are fully aware that this argument is convincing only for scholars who accept that research is never neutral, is always connected to society, and therefore has to be critical and reflective.

Conceptual approaches applied in the case studies of this book, including network analysis, discourse analysis, historical and ethnographic case studies, and social ecology, will be briefly presented. The final part of this chapter provides an overview of the case studies within this book, which illustrate not only new approaches but also the multidisciplinary, complementary, and collaborative kinds of research we believe are necessary.

Present Transformations toward a Glocalized World

At the end of the millennium, "globalization" has attained a prominence within public discourse. Upon close examination, however, the current processes might be better captured by the dialectic term "glocalization" (Robertson 1995). Glocalization, a combined notion of globalization and localization (Robertson 1998, 197), refers to the fact that the current explosion of interterritorial linkages and communications is not just a phenomenon of increased "horizontal" interaction, but also has to be understood in its "vertical" dimension, characterized by direct mergers of local and global processes.

What "glocalization" contributes is a recognition of the greater importance of the local and global levels compared with the interposed national level. Even more important is the shift of emphasis from units, entities, or actors toward the flows, interactions, linkages, and bonds among these units.

In the following paragraphs we will briefly delineate some of the most important aspects of the process of glocalization in several dimensions, starting with the most important economic sphere, followed by the social and political spheres.

The globalization of the economy is driven by increasing international trade, but even more so by the explosion of transnational financial transactions. Facilitated, in 1947, by the negotiated agreements of the General Agreement on Tariffs and Trade (GATT), international trade has grown steadily since World War II. However, the most dramatic processes of economic globalization have been triggered by the dematerialization and symbolization of the economy.

The transformation toward a postindustrial economy is characterized by the relative decline of the first two sectors of the economy (agriculture and industrial production) in comparison to a third: trade and transport. Provision of information and communication has displaced the production of material goods as the primary economic activity in Organization for Economic Cooperation and Development (OECD) countries. As the economy has shifted away from the material toward the symbolic, deregulation and new information technologies have created a global financial system that has effectively debordered the world of nation-state control (O'Brien 1992). Starting with the "Eurodollar market" in London in the 1960s, the world has witnessed an explosion of "offshore markets," as in the Bahamas, for example. Unencumbered by state control, these financial centers process $1 trillion in transactions daily, an amount roughly equivalent to the annual gross domestic product (GDP) of France or double the total currency reserves of all OECD states. Speculation accounts for the overwhelming majority of these financial transactions; only 3–4% are induced by material trade (Neyer 1996, 69).[2]

Whereas some observers interpret these trends toward a dematerialized and globalized economy as "the end of geography" (O'Brien 1992), many others remain skeptical. Proponents of "geopolitics" counter that many conflicts revolve around natural resources, as evidenced within the vast geographic area of the former Soviet Union. Localization matters very much in the postindustrial economy (Krugman 1991). The centers of the new information economy are concentrated within narrow districts of a few global cities. The financial district in lower Manhattan is an especially prominent example. Similarly, the most important companies in the advertising industry are based in a single neighborhood in London. Regional "clusters" of many corporations (e.g., Silicon Valley) are common in many high-tech sectors, and the nurturing of such innovative milieus has proven an indispensable strategy for regional economic development (Krumbein et al. 1994; Storper and Harrison 1991; Maillat et al. 1995). Management guru Kenichi Ohmae (1993) has averred that the consequence of this new economic logic is "the rise of the region state."

Mutual exchange based on a utilitarian rationale is not the sole motive that brings together people across the globe. Transnational interaction and cooperation by social, nongovernmental actors can also be based on shared values and beliefs. The number of international nongovernmental organizations (INGOs) has exploded; as of 1999 there were just over 17,000 organizations like the World Council of Churches, Friends of the Earth, the International Olympic Committee, and the International Federation of Red Cross and Red Crescent Societies (Union of International Associations 1999–2000). Other prominent examples of new global players united by shared values and beliefs are transnational nongovernmental organizations like Greenpeace and Amnesty International.

Some "nonterritorial communities" (Elkins 1995) are based less on shared values and more on shared knowledge or shared interest in a specific theme. For example, the scientific communities of researchers and journalists are groupings that are increasingly bound together on a transnational level through meetings and affiliations like the Association of Environmental Journalists. These transnational networks of knowledge creators and brokers are becoming more important in a world characterized by enormous complexity and information overload. It is not certain, however, that these actors always play a positive role in terms of accountability, as is sometimes assumed by research on "epistemic communities" (Haas 1989, 1992).

Political actors have not merely reacted to these economic and social currents; they have deliberately, or perhaps unwittingly, initiated and institutionalized them (Neyer 1995; Kapstein 1994). Thus we have witnessed the proliferation of international organizations and international issue-based legal regimes since World War II. The roster of more than 3,000 international governmental organizations (IGOs) now dwarfs the number of nation-states. Although most IGOs focus on economic or trade issues, scholars have nonetheless identified more than 250 regional environmental instruments created by 1992 (DiMento, this volume), representing virtually all regions.

Several characteristics of international regimes account for their transformative nature. Most importantly, national members are bound by decisions reached at the supranational level. Some regimes go further, allowing member states or even individual citizens to seek redress in the judicial arena or other forum for dispute resolution. Sanctions may be levied against member states for noncompliance or other violations of regional requirements.

This proliferation of political actors and institutions "above" the nation-state is just one part of the overall picture. Many signs of fragmentation and decentralization of political power have accompanied the movement toward globalization. Since World War II the number of nation-states has more than doubled because of the collapse of empires: Recall the last wave of sovereign states founded amid the rubble of the Soviet Union (Waters 1995, 114). Even more important is a general trend in almost all states toward decentralization and the empowerment of the subnational, regional level during the last fifteen years. Multilevel governance (Marks et al. 1995), federalism and confederal arrangements (Elazar 1998), and subsidiarity are restructuring the architecture of governance all over the world.

Paradoxically, at the same time as institutions and associations with universal values, beliefs, and knowledge bases embrace globalization, we are witnessing a resurgence of cultural, ethnic, and religious sectionalism and particularism. The Islamic "jihad" and reference to "Asian values" are two prominent examples of cultural resistance to the universalistic or, as others argue, imperialistic claims of the West (Barber 1995).

This cultural particularism is apparent not just on a global scale but within sovereign nation-states as well. Regionalism, provincialism, and ethnic nationalism[3] need not proceed as far as separatist movements in

Quebec or the Basque region to undermine the cohesion of the nation-state and challenge the role of the central government as the sole actor in the international and transboundary realms (Duchacek 1986).

It is not just that the locus of decision making is moving "upward" and "downward" from the level of the nation-state. There are many new links and alliances bypassing the central state, a situation Duchacek dubbed "perforated sovereignty" (Duchacek 1986). Cities and other subnational political units like states, Laender, provinces, and cantons are "going global" in their attempts to attract investment and host major events like the world fair EXPO (Michelmann and Soldatos 1990; Brown and Fry 1993). Furthermore, subnational units have fought successfully to have a voice in international affairs, formerly deemed an exclusive domain of the central government. German Bundeslaender possess the right to lead the German delegations in negotiations in Brussels in policy fields that are their responsibility according to the national constitution. This is just one example of a process Brian Hocking (1993) called the "localizing of foreign policy." All these various developments toward a world of intermestic politics (Manning 1977) evince that nation-states are "loosing" their hegemony as units of reference for identities, interest aggregation, and loyalties, as well as their roles as gatekeepers in a "two-level game" (Putnam 1988) between the international and the domestic realm. The nation-state is far from disappearing from the scene, however. Decisions now involve many more actors, and the nation-state often represents a veto point that can stall agreements or derail their implementation.

The most important consequence of the emergence of the glocalized world, as compared to that of the modern era dominated by the reign of nation-states, is the broader range of actors involved. Furthermore, multiple, overlapping memberships between these new actors make distinguishing national affiliation more difficult. Surveying the divergent interests of "American" multinational corporations and the American state and people, Robert Reich (Reich et al. 1990) conceded the difficulty of determining "who is us?" To understand a political conflict over a transboundary watercourse, for example, it is no longer adequate to model a bargaining game between the various riparian or littoral states based on their national interests. The unfolding of the whole range of involved actors, their identities, interests, and loyalties as well as the linkages and alliances between these actors is essential for better understanding and explanation.

Further Transformations beyond Modernity

The sovereign nation-state is not the only aspect of the modern world challenged by the present transformations. Other core elements of the modernity project, such as basic assumptions about knowledge, are likewise confronted by new postmodern and renewed premodern concepts. We will briefly delineate the different features of these elements in respect to their ontology; point to their dominant expression in the fields of technology, economy, law, and politics; and show their correspondence to three different theoretical strands in the political science subdiscipline of international relations.

Figure 1.1 shows the transformations of ontological bases over time. Ontologies, basic assumptions about the world that guide our reasoning and theory building, were first fundamentally transformed during the Enlightenment, resulting in the liberation of mankind from the natural and traditional environment. Within premodern thought humans were considered an integral component of one objective real world. Fate was determined by forces external to the individual, practically by the unforgiving natural environment and philosophically by the rigid doctrines of religion. Human behavior was seen as a reaction and adaptation to these external structures. Characteristics generally attributed to the premodern era are craft as the dominant technical feature, subsistence economy, recourse to "natural law" or to "religion" as normative guidelines, and the monarchy (hierarchical direction from above) as the dominant political concept.

In international relations theory, the realist school conceptualizes a world system consistent with this ontology. State behavior is seen as externally determined by the harsh, anarchic international environment. Because all other actors are considered potential adversaries, states have to act in a competitive manner. Accordingly, the relative-gains principle

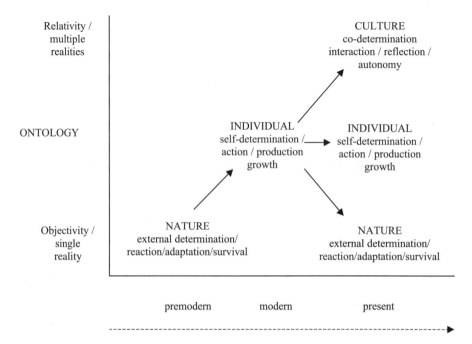

Figure 1.1
Transformations of ontologies over time

governs diplomacy. States strike deals or participate in agreements only when they stand to benefit more than the other side. Power is based on physical (military) capabilities (Waltz 1986).

The onset of the modern era has fundamentally transformed ways of living and thinking. The Enlightenment philosophically liberated the human being from external determinism and furthermore reimagined the individual as the center of the objective universe. "Agency" displaced "structure" as the primary point of reference for conceptualizing the world and constructing theories. Complementing the priority accorded the individual, modern thought also celebrates the notions of self-determination and human control of the natural environment. Thus society has moved from adapting to natural imperatives to manipulating natural resources with the help of industrial technology to produce useful goods. This has resulted in new features of living and reasoning: machines as the dominant technical feature, the industrial economy, the emergence of universal principles based on human rights, and representative democratic governance within territorially defined units (formal hierarchy with "bottom-up" directions) as the dominant political concept.

Modern reasoning corresponds to the liberal approach in international relations theory (Moravscik 1997). This approach privileges the preference-building process within nation-states, thus conceptualizing the international arena as a second level where states bargain based on their domestic preferences. States act in a "selfish" or "individualistic" manner in the sense that they are looking only at their own gains and losses. The point of reference for comparison, unlike in the relative-gains approach, is not the other actors but the present state of affairs or potential outcomes from alternative courses of action. Power is based on economic strength.

Modern concepts have never totally transformed the world and its societies, but the modernity project has presented a radical alternative to the premodern paradigm. What makes the present transformation confusing and puzzling is the fact that we are witnessing simultaneously the resurgence of premodern elements, the ongoing existence of modern concepts, and the emergence of postmodern characteristics (see figure 1.1).

At the beginning of the twenty-first century, we are confronted with novel ways of conceptualizing the world. A new ontology is emerging that is pushing the concept of reality even more toward pluralism and relativism. The French poststructuralists (Baudrillard, Derrida, and Lyotard) have contended that our understandings of the world are socially constructed. If our assumptions are inevitably shaped by our language, which is also a product of social interaction and varies by culture, there can exist no single objective reality. Furthermore, it is impossible to ascertain the superiority of one reality over another (Gandy 1996).

The core of the new ontology is the notion of culture as socially constructed. Human behavior is seen neither as externally determined by an anonymous environment nor as self-determined according to individual preferences, but as codetermined by social interdependence and interaction. The present variety and relativity of cultures, values, and realities leads to a situation in which constructing and defining personal identity, rather than adapting to the natural environment or producing material goods, becomes the primary task for human beings. Many societal transformations have accompanied this emerging ontological base: dematerialization of production (microelectronics and genetics as new core technologies), a service- and information-based economy, the tolerance of various value sets based on the concept of multiculturalism, and multiple, overlapping units for collective action and identification based on various kinds of ties (territory, ethnicity, belief, taste, sex, among others).

These features and the relativistic ontology are reflected in the constructivist strand in international relations theory. When conceptualizing the behavior of political actors, constructivists focus their attention on normative-cognitive elements of that behavior such as the identities of political actors, problem definitions, perceptions, communication, and shared understanding. Power is based on network centrality and communication skills (Kubalkova, Onuf, and Kowert 1998; Adler 1997).

Abundant evidence exists that all of these ontologies—despite their contradictions—will be useful foundations for describing, explaining, understanding and *creating the present and future world* (we will provide such evidence in the field of transboundary water policy in chapter 2).[4]

There are many intriguing questions about the relationship between the various ontological bases and their corresponding worldviews. What is the relationship between nature and culture? Many environmentalists have brandished natural and cultural arguments simultaneously in their fight against products of modern human megalomania. In fact, they have conflated culture and nature by assuming that traditional cultures embody the knowledge needed to coexist peacefully with the natural environment. However, culture is now conceptualized as the shared normative-cognitive beliefs, or worldviews, of a social community, rather than the accreted sediment of previous experience.

We can differentiate cultures based on local spaces (spaces of places) and time past (experience, tradition), from cultures based on global spaces (spaces of flows) and present and future times (discourses, visions). Liberating culture conceptually from the past raises more questions: Can present discourses and visions about the future exist independently of past experiences? Can cultural inventions overcome any natural limits?

Confusingly, the equalization of nature and culture is also challenged the other way around. As David Hughes shows in chapter 10, traditional cultures can stand at odds with new ecosystem concepts like watershed planning. In this example, the culture is bound to the past and the concept that is based on a "natural" imperative is new. From this perspective we encounter an equally puzzling question: How much are natural or ecological imperatives really based on objective realities and how much are they creations of the human imagination? The idea of "bioregions" as primary territorial units for governmental planning and control, discussed in chapter 5 by Suzanne Levesque, is an example well suited to discussion of this question.

An ability to tolerate ambiguity is requisite for individuals who live in this complex world of multiple and contradictory realities, not to mention researchers who try to understand and to explain these realities. The basic goal of this book is to facilitate and encourage emerging ways of thinking about transboundary water and natural resources. It is not that such approaches are new or novel, but rather that they have been generally marginalized in the study of water policy. Perspectives once relegated to the practice squad are, in this book, afforded significant playing time.

Emerging Approaches for Studying Transboundary Water Policy

The modern search for generalized rules and human control has led to scholarship in which predictability, parsimony, and simplicity have been the measure of academically acceptable approaches and methodologies. These precepts have often led to single-disciplinary research on specific and fairly narrow research topics. In the face of the increasing variety and complexity of the world in which we live, in which many forces are often interacting from different directions, singularity of focus and false parsimony of theoretical concepts oversimplify to the point of being misleading.

Emerging approaches that encourage the introduction of various disciplinary perspectives and their differing methodologies are far more useful in capturing variety and complexity. To capture the realities described above, it is not necessary to integrate disciplines into interdisciplinary systems analysis and other metatheoretical frameworks. More insight often emerges from a panoply of studies, each of which delves deeply into the aspects of problems using the disciplinary tools best designed to examine the specific problem. Research in such approaches is like a tapestry, each strand of which may be quite separate and distinct but not really significant until it is combined with others into a whole within which patterns emerge.

The most recent decade has witnessed the emergence of a number of new theoretical approaches and methods useful for the study of water. We will concentrate within this book on applications and extensions of modern approaches in a more relativistic direction. Whereas the life sciences have been widely accepted as a necessary supplement to the fields of water and natural resources policy, no similar welcome has been extended to the social sciences and humanities. It is time to make progress in this direction as well. A second reason for such approaches has been stated before: We believe that these approaches, which concentrate on understanding the involved meanings of water, have to be the first step in every progressive research project. This implies that there is no definite connection between a specific meaning of water and a specific method.

The research methodologies introduced here—network analysis, discourse analysis, ethnographic and in-depth historical case studies, social ecology, and the ways of knowing that these reflect—are only some of the emerging analytical approaches to the study of water. Others, such as participatory research, environmental mediation, and practical peacemaking, are becoming more important, have clear implications for democratizing research, and directly link thought to action. The research methods we have chosen to describe in this chapter, however, are the ones reflected in the subsequent chapters of this book. These approaches can enhance analytical and applied efforts to understand transboundary water problems.

Game theory is not included in this book even though it is one of the most sophisticated analytical tools in contemporary social science. Game theory is an analytical advance in that it moves from action-based ontology toward an interaction-based ontology. It uses the advanced modern tool kit of mathematics. Although we acknowledge game theory's potential for explaining many conflicts and outcomes in current cross-border environmental politics, we believe it is not as helpful as the new approaches advanced here in the messy world in which we find ourselves. The glocalized world is characterized by a multiplicity of actors and the permanent construction and reconstruction of identities, boundaries, and preferences of these actors. The "logic of consequentiality" (March and Olson 1989) is less and less appropriate as a successful strategy in a "debordered," "hyperconnected," and turbulent world.

An overview of the focus of each of these emerging analytical frameworks is provided below.

Network Analysis

In many parts of the world, subnational actors are beginning to drift away from their intranational moorings and embark on a more transnational pattern of activity. This shift, in conjunction with the omnipresence of organized actors in policymaking, the overcrowding of participatory arenas, the fragmentation of the state, and the blurring of boundaries between the public and the private, has created an environment in which networks emerge as a policy-relevant analytical unit (Kenis and Schneider 1991, 41; Mayntz 1993). Network analysis provides a framework and a set of techniques for studying relationships within and across groups of social actors. It focuses on communication patterns among individuals to understand the dynamics within and between groups and networks of people. The methodology of network analysis identifies linkages and network boundaries, the understanding of which is useful in explaining certain policy outcomes. Network analysis as an analytical tool has developed sophisticated formal methods. Most network analysts base their approach on a rationalistic action theory (strategic exchange of resources by interdependent actors). There are also conceptualizations more relevant to this book that stress the normative-cognitive factors that guide the clustering of political actors in policy processes. Paul Sabatier's "advocacy coalition frameworks" (ACFs) and Ernst Haas's "epistemic communities" are prominent examples of these more relevant conceptualizations. In Sabatier's and Haas's approaches, shared beliefs, and not resource dependencies, are the central focal points that bind actors together.

A central concept of ACFs is the "advocacy coalition," which is "composed of people from various government and private organizations who both (a) share a set of normative and causal beliefs and (b) engage in a non-trivial degree of coordinated activity over time" (i.e., seven to ten years) within a particular policy subsystem (Sabatier and Jenkins-Smith 1999). Members of an advocacy coalition share beliefs on topics such as human nature, prioritization of values (e.g., freedom, power, and beauty), and justice.

Haas defines an "epistemic community" as "a specific community of experts sharing a belief in a common set of cause-and-effect relationships as well as common values to which policies governing these relationships will be applied" (Haas 1989, 384). According to Haas, the primary function of these epistemic communities is reducing uncertainty. These networks of experts are able to find a common denominator with respect to definition of problems and general strategies to solve these problems in policy fields that are characterized by high complexity and uncertainty.

Discourse Analysis

Networks of people rely on a shared discursive framework to communicate with each other. Embedded in each discursive framework is a set of normative values and preferred modes of reaching agreement. Discourse analysis elucidates how to identify the dominance of one way of knowing within a group of people in a policy environment. People sharing an interest in solving a common problem often end up talking past each other; discourse analysis facilitates understanding of why this occurs. Discourse analysis facilitates a recognition that differences in frames of reference, interpretations, and meaning are at the root of many misunderstandings.

Following the lead of such scholars as Habermas (1971) and Foucault (1977), much recent scholarship has focused on the contents of communications and the ideas embedded in discourse. This kind of analysis recognizes that language and symbols determine how debate is framed, what issues are placed on political agendas for discussion, and what alternatives are likely to be viewed as viable. Much of this kind of analysis is explicitly critical and intends to be transformational.

Discourse analysis brings to ways of understanding (and shaping) our world a recognition of the powerful role that the social construction of meaning plays in our lives and communities. Postmodern discourse analysis illuminates various competing systems of language or discourses and allows us to talk about why some discourses (like those concerning progress, technocratic rationality, and universal rights) are more convincing and more authoritative than others.

Historical and Ethnographic Analysis

With the growing recognition that general laws, once held as the highest goal of science, embody biases in their assumptions and hold for very few actual human situations, researchers have returned with renewed interest to deeply textured and carefully documented historical case studies and ethnographic analysis. More than those that result from other kinds of analysis, these studies allow people to speak directly through personal diaries, oral histories, and family records as well as through official documents that have long been the stock and trade of historians. Careful historical documentation of land ownership patterns, movement of peoples, and the development of superior/subordinate relationships among them help scholars understand, for example, why animosities persist, why practices that seem irrational continue, and why some present-day conflicts are so difficult to resolve.

Historical and ethnographical analysis accords context a central place in explanations and understandings of specific phenomena. A context of rapid social and economic change provides different opportunities and challenges than does one of stability. Definitions and social constructions that take the form of immutable laws at some times and in some cultures may be highly contested and in flux in others. Institutions as well as peoples and nations have different cultures that affect processes and actions but are not obvious unless practices are observed historically.

The creation of boundaries and the occurrence of cooperation and conflict across them cannot be understood without specific, locally based knowledge of identities, stereotypes, and social constructions of tribal, racial, and ethnic groups, the roots of which may be best illuminated through ethnographic and historical analyses. Approaches involving these types of analyses are also able to identify which impulses are likely to be acted upon in specific situations and which other attitudes cause acquiescence to prevailing policy on the part of locally based tribal, racial, and ethnic groups.

Social Ecology

Like discourse analysis, social ecology allows for and facilitates communication across different ways of knowing. A number of authors in this volume (Levesque, Doughman, DiMento, Garb, Ingram, Perry, and Whiteley) are associated with the School of Social Ecology at the University of California at Irvine and reflect its perspectives in their work. Key principles of social ecology are systems thinking, contextual research and theory development, interdisciplinarity, and community problem solving. Consistent with this view, social ecology employs quantitative and qualitative data analysis, both hypothesis testing and grounded theory research approaches, reductionist and integrative design, and positivist and postmodern discourses.

In all its manifestations, social ecology is rooted to place; social ecologists view the local, physical environment as an important component of understanding and analyzing sociopolitical contexts. Social ecological approaches often characterize human-environment interactions as two-way or cyclical systems in which, for example, human action affects environmental health and environmental health affects human health. Part of this commitment to the local context stems from a belief that academic research should facilitate understanding and amelioration of recognizable and pressing social problems.

Organization of the Book

The chapter that follows in this first, introductory section of the book continues the discussion begun in this chapter of the need for more innovative and diverse frameworks and methodologies. In chapter 2, the central subject is the inadequacy of standard legal, engineering, and economic frameworks for capturing the value-laden and symbolic attributes with which water is endowed at the end of the twentieth century. We demonstrate that the "disembedded" modern conceptualizations of water as product, as property, and as commodity have to be complemented by notions of water, like "gift of nature," "focal point for community building," "security issue," or "specific good," that connect water to the natural and cultural environment. Our pointing to these meanings does not reflect an intent to promote them as normatively superior but rather to foster the acknowledgment of their very existence.

The subsequent eight chapters contain case studies that make different use of the frameworks laid out in the introductory chapters. The first case, discussed by María Rosa García-Acevedo in chapter 3, fulfills the goal established in this chapter of understanding the meaning(s) of water in specific contexts. García-Acevedo traces a broad range of meanings of water as they emerged chronologically, thereby laying the groundwork for more specific studies. In contrast to her more comprehensive treatment, other case studies throughout the book highlight one or two meanings of water. Whereas García-Acevedo's chapter shows mainly the change from premodern to modern meanings of water, the subsequent two chapters clearly show the growing importance of postmodern phenomena in transboundary water politics. Joachim Blatter and Suzanne Levesque show, in chapters 4 and 5, the importance of cross-border advocacy coalitions based on shared belief systems and postnational ideas of governance in Europe and North America. Blatter openly demonstrates the superiority of an interpretative approach to explain cross-border water policy over functionalistic and rationalistic approaches. Levesque supports the postmodernist perspective in her concentration on transboundary networks based on socially constructed meanings of water.

The next two pairs of case studies challenge Blatter's and Levesque's postmodernist approaches on two different levels: The first two, by Sullivan (chapter 6) and Doughman (chapter 7), use the postmodern tools of discourse analysis but take issue with the optimism and progressivism of the case studies in chapters 4 and 5. The second two, by Garb and Whiteley (chapter 8) and DiMento (chapter 9), challenge postmodernism by insisting that modern approaches are still quite useful in many parts of the world, at least insofar as capturing the fundamentals of a particular case are concerned. The postmodern influences in these two cases appear significant but marginal. The case study presented by David Hughes in chapter 10 is included as the closing case study because it makes comprehensive use of the overall framework and thus parallels the opening case study in chapter 3. At the same time, it shares much of the skepticism of the two case studies that precede it. Hughes's case study demonstrates very impressively that different meanings of water can coexist at the same time, in contrast to the more evolutionary description in chapter 3. Moreover, chapter 10 provides us with the opportunity to debate a fundamental challenge to the usefulness of an analytical framework based on Western modernization theory in the concluding sections of the book, chapters 11 and 12, by Perry, Blatter, and Ingram.

Summary of Chapters

Chapter 2, by Joachim Blatter, Helen Ingram, and Suzanne Levesque, catalogues in detail the very diverse and symbol-laden meanings of water that have emerged at the beginning of the new millenium. The chapter broadens from narrow modernist constructions of water as property, product, and commodity to include older, premodern meanings of water as a gift of nature as well as more recently emergent postmodern meanings.

In her case study entitled "The Confluence of Water, Patterns of Settlement, and Constructions of the Border in the Imperial and Mexicali Valleys (1900–1999)", María Rosa García-Acevedo discusses the ways water, the nature of boundary and settlement patterns, and the culture and livelihood of people are related. Through contextual historical analysis that relies on sources written in Spanish as well as English, the author identifies the changing meaning of water as a central variable. The chapter recounts the dispersion and marginalization of indigenous peoples; the uneven agricultural development of valleys on either side of the national dividing line between Mexico and the United States that otherwise appear potentially equally productive; and the lopsided pattern of industrialization that currently undergirds the population explosion in this arid area. The varying roles of national governments, settlers' groups, and commercial entrepreneurs are traced through time. Distinct differences among winners and losers in terms of control over and access to water and water quality are identified. García-Acevedo concludes that the capture of the definition and meaning of water is a highly political and value-laden process.

A focus upon transnational networks, particularly advocacy coalitions, ties chapter 3, by Joachim Blatter, entitled "Lessons from Lake Constance: Institutions, Ideas, and Advocacy Coalitions," to the chapter that follows it. Lake Constance represents one of the most successful cross-border water protection regimes in the world. To obtain a better understanding of the preconditions of this success, Blatter conducts an in-depth study focusing on the regulation of motorboats on the lake. His findings demonstrate that when modern explanations for analyzing regulations fail, only a recourse to history and discourse provides basic insights into policy preferences and outcomes related to motorized boating on the lake. Blatter further demonstrates that territorial aggregation of interests and a nation-state-centered approach no longer capture the reality of transboundary politics as it occurs on the Lake. Finally, Blatter's case study shows an interesting complementarity between the formal international commissions and the informal cross-border regional networks that have emerged as responses to the continental integration process in Europe.

Chapter 5, "The Yellowstone to Yukon Conservation Initiative: Reconstructing Boundaries, Biodiversity, and Beliefs," by Suzanne Lorton Levesque, presents an analysis of the Yellowstone to Yukon Conservation Initiative (Y2Y), a transnational grassroots environmental network that has emerged and is attempting to influence environmental decision processes in the binational region embodied in its name. In this region socioeconomic structures and environmental perspectives are changing rapidly. New actors are emerging to challenge historic constructions of nature and resources and to demand a voice in environmental decisions. Levesque focuses on Y2Y's extensive use of information and communication technologies to create a nature-centered discourse, to inform and broaden the network, to achieve consensus on problem definitions and strategy preferences, to elicit broad-based support from the public, and to pressure government and economic decision makers.

Chapter 6, "Discursive Practices and Competing Discourses in the Governance of Wild North American Pacific Salmon Resources," by Kate Sullivan, concentrates on public-sphere discourse regarding water and fisheries. Using the case of the Pacific Salmon Treaty, Sullivan examines the print-mediated public debates over the governance of wild North

American Pacific salmon resources in British Columbia and Washington. Complexly interlaced discourses about biological sustainability and economic rationality stimulate vigorous debates over economic equity, allocation, conservation, and resource uses. The chapter considers the role of the media, technocratic-rational language, and the groundswell of grassroots coalitions of stakeholders in the discursive practices and framing discourses mobilized in the very contentious and high profile United States–Canadian Pacific Salmon Treaty negotiations. The chapter reveals that language and conceptual frameworks have very concrete material outcomes for both economic well-being and environmental health.

In chapter 7, "Discourses and Water in the U.S.-Mexico Border Region," Pam Doughman employs discourse analysis in a very different way from the preceding chapter. Doughman critically examines the discourse used in the formal documents and informal debates and communications of two new institutions involved in managing water resources in the U.S.-Mexico border region, the North American Development Bank (NADBank) and the Border Environment Cooperation Commission (BECC). These institutions were established under one of the environmental side agreements of the North American Free Trade Agreement as a means of increasing procedural democracy in transboundary water management. Like Sullivan, Doughman is interested in the extent to which public-sphere discourse is inclusive and representative of citizens' points of view. In her chapter, Doughman asks whether the institutional perspective of the NADBank and BECC differs significantly, in terms of inclusiveness and representation, from the managerialist perspective that dominates the International Boundary and Water Commission.

Chapter 8, "A Hydroelectric Power Complex on Both Sides of a War: Potential Weapon or Peace Incentive?" by Paula Garb and John Whiteley, discusses the joint management of a hydroelectric power plant by two warring parties in the South Caucasus. In the 1970s, the plant was built on both sides of the Inguri River in the former Soviet republic of Georgia. No one could have predicted then that the power plant might someday be situated on two sides of a disputed border policed by international forces. In 1992–93, when war was being waged between the newly independent Georgian government and the separatist government of the former autonomous republic of Abkhazia within Georgia, both sides depended on electricity from the Inguri plant, and neither side could run the power plant without cooperation from the other side. The only way to keep the plant operating and electricity generating was for the governments to agree on the terms of joint management and financing, even though they could not agree on the terms of a peace settlement. The chapter provides insights into the factors that have facilitated cooperation between the leadership of the respective energy agencies and between the plant employees on both sides of the dispute.

The findings of Joe DiMento in chapter 9, "Black Sea Environmental Management: Prospects for New Paradigms in Transnational Contexts," present a complementary focus on the supranational regimes that instigate the subnational networks revealed as so important in the Blatter study. In this chapter DiMento sets forth the history of the Black Sea Environmental Program, which focuses on the need to address monumental social, political, and environmental problems after the collapse of the Soviet Union. Formal international organizations such as the United Nations Environmental Program and the Global Environmental Facility are the catalyst for the creation of a faltering but nonetheless important effort at international cooperation in an area where nations are distinctly different and have many reasons for conflict. Characteristic of social ecology research, DiMento weaves physical description of the state of the endangered Black Sea through legal and social commentary.

Chapter 10, "Water as a Boundary: National Parks, Rivers and the Politics of Demarcation in Chimanimani, Zimbabwe," by David Hughes, is an in-depth, place-based case study concerned with the erection of water-related borders within regions. Hughes focuses on territorial conflict related to "linear" water-marking property lines, which has a very different meaning from the number of more volumetric definitions explored in the Imperial/Mexicali Valleys case. Hughes maintains that water is inseparable from land and the patterns of property and authority that govern land. His chapter explores the interdependency of water and territory in the context of a long-running dispute over the border of the Chimanimani National Park in eastern Zimbabwe.

The two final chapters in this book mirror the two introductory chapters. Having started with the general transformations toward a glocalized world and their implications for the analysis of transboundary water policy, we end with an attempt to transfer the lessons learned from new investigations into transboundary water politics toward more general considerations of the future of governance and social science.

Before we return to our starting point, however, Richard Perry, in chapter 11, "Perspectives from the Districts of Water and Power: A Report on Flows," turns the thesis of the book on its head. The introductory section of the book maintains that in the contemporary globalized world, transboundary water policy requires new forms of analysis that go beyond modernist frames of thought: engineering, law, and economics. Perry argues that water itself is a good way to think about present global transformations. He traces how water and the emergence of the nation-state were inextricably intertwined and demonstrates how the image of flowing water aptly captures the contemporary fluidity of capital and power. Perry exemplifies the way in which discursive framing enhances comprehension. Perry employs water symbols and images to create an understanding of both historic and present practices that could not be captured in a more technical or modernist language.

Finally, in chapter 12, "Lessons from the Spaces of Unbound Water for Research and Governance in a Glocalized World," Perry, Blatter, and Ingram revisit the eight case studies in chapters 3–10, drawing conclusions about them based on the theoretical framework developed in the first two chapters. They demonstrate the usefulness of the general framework presented in chapters 1 and 2, but not without self-reflection and self-criticism. Such conscious self-criticism is, in our view, essential for present-day transboundary water researchers. Much of the previous modern research we wish to augment has been flawed by the imperialism of Western modes of thought to which it is easy to fall prey. Therefore, it is appropriate to conclude with a discussion that challenges the theoretical and methodological underpinnings of the book. This should not be interpreted as a signal of the authors' weakness or lack of belief in the usefulness of this endeavor. Rather, we are simply taking to heart a central tenet of the book: the current transformational state of the world demands self-conscious reflection about the fundamental ontological and epistemological bases of our approaches to govern, to manage, and to study all critical issues, most especially water.

Notes

1. Giddens (1990) has labeled this process "disembedding."

2. Neyer (1995, 69) cites management authority Peter F. Drucker: "In the world economy of today, the 'real' economy of goods and services and the 'symbol' economy of money, credit and capital are no longer bound tightly to each other; they are, indeed, moving further and further apart."

3. Because most modern "nation-states" comprise many nations defined as ethnically homogeneous groups, this kind of new "postmodern" nationalism is targeted against the modern nation-state (Barber 1995, 165).

4. Note that we are not advocating these new ontologies because we think that the world is necessarily improved by these transformations, but because we think they are necessary to understand the present and future world. Better understanding is a necessary but not a sufficient precondition for better practice.

References

Adler, Emanuel. 1997. "Seizing the Middle Ground: Constructivism in World Politics." *European Journal of International Relations* 3, no. 3: 319–363.

Barber, Benjamin. 1995. *Jihad vs. McWorld*. New York: Random House.

Brown, Douglas M., and Earl H. Fry (eds.). 1993. *States and Provinces in the International Economy*. Berkeley: Institute of Government Studies Press, University of California.

Duchacek, Ivo D. 1986. *The Territorial Dimension of Politics within, among, and across Nations*. Boulder, Colo., and London: Westview Press.

Elazar, Daniel J. 1998. *Constitutionalizing Globalization: The Postmodern Revival of Confederal Arrangements*. Lanham, Md.: Rowman & Littlefield.

Elkins, David J. 1995. *Beyond Sovereignty: Territory and Political Economy in the Twenty-First Century*. Toronto: University of Toronto Press.

Foucault, Michel. 1977. *Bewachen und Strafen: die Geburt des Gefaengnisses*. Frankfurt am Main: Suhrkamp.

Gandy, Matthew. 1996. "Crumbling Land: The Postmodernity Debate and the Analysis of Environmental Problems." *Progress in Human Geography* 20, no. 1: 23–40.

Giddens, Anthony. 1990. *The Consequences of Modernity*. Stanford: Stanford University Press.

Haas, Peter M. 1989. "Do Regimes Matter? Epistemic Communities and Mediterranean Pollution Control." *International Organisation* 43, no. 3 (Summer): 377–403.

Haas, Peter M. 1992. "Introduction: Epistemic Communities and International Policy Coordination." *International Organisation* 46, no. 1 (Winter): 1–35.

Habermas, Jurgen. 1971. *Toward a Rational Society*. London: Heinemann.

Hocking, Brian. 1993. *Localizing Foreign Policy: Non-central Governments and Multilayered Diplomacy*. London and New York: Macmillan.

Kapstein, Ethan B. 1994. *Governing the Global Economy: International Finance and the State*. Cambridge: Harvard University Press.

Kenis, Patrick, and Volker Schneider. 1991. "Policy Networks and Policy Analysis: Scrutinizing a New Analytical Toolbox." In Bernd Marin and Renate Mayntz (eds.), *Policy Networks*, pp. 25–59. Frankfurt: Campus.

Krugman, Paul. 1991. "Increasing Returns and Economic Geography." *Journal of Political Economy* 99, no. 3: 483–499.

Krumbein, W. E., P. Brimblecone, D. E. Cosgrove, and S. Staniforth. (eds.). 1994. *Durability and Change: The Science, Responsibility, and Cost of Sustaining Cultural Heritage*. Chichester, N.Y.: John Wiley.

Kubalkova, Vendulka, Nicolas Onuf & Paul Kowert (eds.). 1998. *International Relations in a Constructed World*. Armonk, N.Y.: Sharpe.

Maillat, Denis, Bruno Legoc, Florian Nemeti, and Mark Pfister. 1995. "Technology District and Innovation: The Case of the Swiss Jura Arc." *Regional Studies* 29, no. 3: 251–263.

Manning, Bayless. 1977. "The Congress, the Executive, and Intermestic Affairs: Three Proposals." *Foreign Affairs* 55: 309–315.

March, James G., and Johan P. Olson. 1989. Rediscovering Institutions—The Organizational Basis of Politics. New York, London: Free Press.

Marks, Gary, Liesbet Hooge, and Kermit Blank. 1995. "European Integration from the 1980s: State Centric v. Multi-level Governance." *Journal of Common Market Studies* 34, no. 3: 341–378.

Mayntz, Renate. 1993. "Policy Netzwerke und die Logik von Verhandlungsystemen." In Adrienne Herities (ed.), *Policy-Analyse; Kritik und Neuorientierung*, PVS-Sonderheft 24/1993, pp. 39–56. Opladen: Westdeutscher Verlag.

Michelmann, Hans J., and Panayotis Soldatos. 1990. *Federalism and International Relations: The Role of Subnational Units*. Oxford: Clarendon Press; New York: Oxford University Press.

Moravscik, Andrew. 1997. "Taking Preferences Seriously: A Liberal Theory of International Politics." *International Organisation* 54, no. 4 (Autumn): 513–553.

Neyer, Jurgen. 1995. "Globaler markt und territorialer Staat: Konturen eines wachsenden Antagonismus." *Zeitschrift fuer internationale Beziehungen* 2, no. 2: 287–315.

Neyer, Jurgen. 1996. *Spiel ohne Grenzen. Weltwirtschaftliche Strukturveraenderungen und das Ende des sozial kompetenten Staates*. Marburg, Germany: Tectum Verlag.

O'Brien, Richard. 1992. *Global Financial Integration: The End of Geography*. London: The Royal Institute of International Affairs.

Ohmae, Kenichi. 1993. "The Rise of the Region State." *Foreign Affairs* 72 (April 1): 78–87.

Popper, Karl. 1966. *The Open Society and Its Enemies.* 2 vols. Princeton: Princeton University Press.

Postel, Sandra. 1999. *Pillar of Sand: Can the Irrigation Miracle Last?* New York: W. W. Norton.

Putnam, Robert D. 1988. "Diplomacy and Domestic Politics: The Logic of Two-Level Games." *International Organisation* 42, no. 3 (Summer): 427–460.

Reich, Robert B., Todd Hixon, and Ranch Kimball. 1990. "Who Is Us? (The Changing American Corporation)." *Harvard Business Review* 68, no. 1 (Jan.–Feb.): 53–64.

Robertson, Roland. 1995. "Glocalization: Time-Space and Homogeneity-Heterogeneity." In M. Featherstone/S. Lash/R. Robertson (eds.), *Global Modernities.* London: Sage.

Robertson, Roland. 1998. "Glokalisierung: Homogenitaet und Heterogenitaet in Raum und Zeit." In Ulrich Beck (ed.), *Perspektiven der Weltgesellschaft*, pp. 192–220. Frankfurt am Main: Suhrkamp.

Sabatier, Paul A., and Hank C. Jenkins-Smith. 1999. "The Advocacy Coalition Framework: An Assessment." In Paul A. Sabatier (ed.), *Theories of the Policy Process*, pp. 117–166. Boulder, Colo.: Westview Press.

Storper, M., and B. Harrison. 1991. "Flexibility, Hierarchy and Regional Development: The Changing Structure of Industrial Production Systems and Their Forms of Governance in the 1990s." *Research Policy* 20 no. 5 (Oct.): 407–422.

Union of International Associations. 1999–2000. *Yearbook of International Organizations, 1999–2000. Vol. 4: Bibliographic Volume. International Organization Bibliography and Resources.* Brussels: Union of International Associations.

Waltz, Kenneth. 1986. "Laws and Theories." In Robert O. Keohane (ed.), *Neorealism and Its Critics*, pp. 27–46. New York: Columbia University Press.

2

Expanding Perspectives on Transboundary Water

Joachim Blatter, Helen Ingram, and Suzanne Lorton Levesque

The intention of this chapter is to contribute to an amplified, robust, and flexible understanding of the meaning of water. Inevitably, previous scholarly work on transboundary water resources has been strongly influenced by the dominant features of the modern world. Woven into the very fabric of our existence has been an unflagging conviction that the mysteries of the universe could be unraveled through human intelligence. Vagaries of nature, such as water and the droughts and floods associated with it, could be subdued and harnessed for the benefit of humankind. The companion modern view of society has been that human management of water resources could be rationalized and controlled through laws, institutions, and organizational structures.

The fluidity of water—its unpredictability, variability, and resistance to control—makes it an appropriate metaphor for the inability of modern convictions and perspectives to capture the unstable and fluctuating world. The transformation of key elements of the modern era, including the present process of glocalization, is mirrored in the changing images of water. Discerning such changes is important, not just because it is instructive for many other issue areas, but also because water is essential to nature and to society.

This chapter will briefly discuss how emerging approaches are adding to, and partially displacing, previously dominant modern approaches that were too narrow and too bound to specific ontologies. It will then move on to discuss how the meanings of water imposed by the three ascendant modern disciplines in the study of water resources—law, engineering, and economics—have been modified and transformed. Finally, it will offer a preview of the meanings of water that will emerge in the case studies that follow this chapter.

Expanding Approaches

Lawyers have defined water as a property of territorial units; in the case of transboundary watercourses, of nation-states. Engineers have treated water as a natural resource transformable into products for human consumption. From an economist's perspective, water is a commodity that can be exchanged and traded between various places and various uses.

Although the transition from premodern to modern perspectives has changed the core assumption of our thinking about water from adaptation to nature to control of nature, the dominance of law, engineering, and economics ensured a narrow, bounded set of meanings of water. The present transformation is characterized by expanded meanings and definitions of water beyond these narrow, bounded meanings. The modern (individualistic, rational, and utilitarian) perception of water must be complemented by other understandings of water. These expanded conceptions of water point in two directions (see figure 2.1): First, a renewed awareness of natural imperatives results from insights from the life sciences, especially from research on ecosystems. Second, a new recognition arises that the meanings of water are multiple and bound to various cultures. This social constructivist perspective is based primarily on insights and discussions in the humanities.

Even though both directions share emphases on interdependence, connectedness, linkages, and context, as opposed to the starkly dichotomous contrast between the premodern (structure) and modern (agency) perspectives, they are antagonistic in their ontological and epistemological bases. Whereas the life sciences assume there exists one "real" world that can be discovered and "explained," the humanities adopt a relativistic worldview and instead attempt to "understand" the various worlds within their specific contexts. Both challenge the belief in human control of water that has been so central to the modern era, but the types of challenges each poses are quite different.

The life sciences remind us that water is bound to a territorial (or natural) context, meaning that water cannot be easily appropriated to territorial units with sharp demarcations (spaces of place, like states) because its bonds to territory follow a different geography (spaces of flows, like habitats of salmon). Further, they assert that water cannot be totally

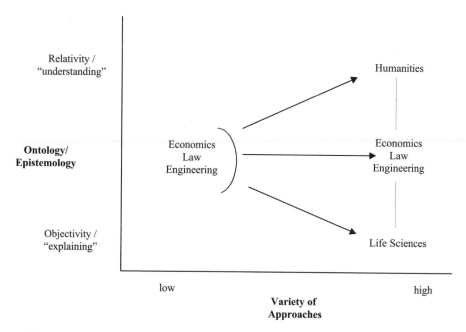

Figure 2.1
Expanding the variety of scientific approaches

controlled, citing the devastating results of many attempts to do so. Finally they show that water cannot be treated as a commensurable commodity because the transfer of water may have serious consequences for both its places of origin and its destination. The ecosystem approach believes that there are objective limitations to human activity, as natural laws resist manipulation.

By noting that water is bound to various cultures, the constructivist approach of the humanities challenges the modern approaches in a much different fashion. Together with the emerging insight that the sovereign, territorially defined nation-state is historically contingent and that other, nonterritorial units of governance are possible, this approach questions the modern canard that territorial units (states) are the only units of reference in the appropriation game. It further highlights the idea that water cannot be treated only as an external product; it is also an essential component of the creation of identities of traditional and new communities. Finally, seen from this perspective, the various meanings and

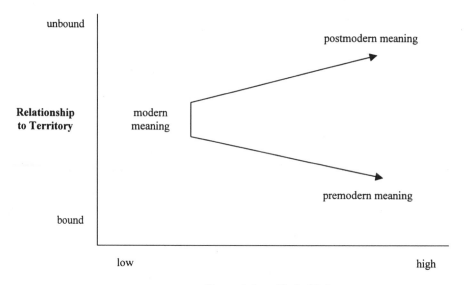

Figure 2.2
New contingencies in the meaning of water

values of water are not easily monetized. Since the various meanings and values of water cannot be conceptualized within one single dimension, the economic rationale for Pareto optimality is stripped of its legitimacy as a "fair" and rational principle for allocation.

Plural Meanings of Water

In response to unresolved equity, environmental, and political problems, a broader understanding of the meanings of water is necessary to overcome the limitations inherent in modernist approaches. Many meanings of water are currently invoked in debates about water policy: Water is a physical/chemical resource; it is a tangible substance. Water is also imbued with various meanings that are socially constructed. Water has economic value; it is a subsistence resource, a component of national security, and a focal point of identity.

Figure 2.2 attempts to disentangle two core elements of this process of transformation and expansion. The horizontal dimension reflects the embeddedness or connectedness of water to social actors and the conceptions of how water is bound to the natural environment. In modern thought, water is regarded as peripheral to social actors: a useful, but not essential, asset. Furthermore, it is perceived as an object that can be disconnected from its natural environment. Water is also currently perceived as a critical element crucial to survival or as a core belief that unites collective actors. Whereas modernity has broken up the unity of subject and object, casting water as a resource to be manipulated and subjugated by engineers, lawyers, and economists, new or renewed meanings seek to return water to the core of existence of political actors. The vertical dimension of figure 2.2 represents the extent to which water is bound to natural territory. At one extreme, water is naturally embedded in place, whereas at the other it becomes a purely socially constructed, artificial, or virtual reality not necessarily bound to any one place.

Figure 2.2 highlights an important insight regarding the current transformation processes. Although new or renewed meanings of water share some common elements that are quite distinct from modern notions, these new meanings also differ sharply from one another. A better understanding of the new contingencies in the meaning of water will follow the disentangling of these (parallel and contradictory) elements. Though modern perceptions of water persist, their hegemony is waning. In the following section, we will examine how the three approaches—legal, technical, and economic—have both expanded and been fundamentally transformed.

From Property toward Essentials: Security and Identity

International law treats water as property, a thing that territorially defined political units can appropriate and own. Yet in many contexts, the construction of water as property obfuscates its importance to the security and identity of states and communities. Figure 2.3 illustrates how the meaning of water varies along the dimensions of importance and territoriality.

On the horizontal dimension, we see the increasing importance of water as we move away from the modern notion of water as property.

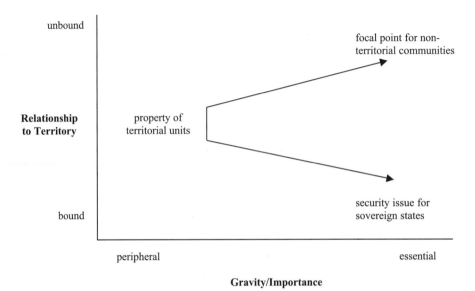

Figure 2.3
Beyond legal approaches: The gravity of water

When understood as simple property, water can have a rather peripheral connection to individuals and social groups, because it is regarded as only one possession among many others. This limited association intensifies as water becomes integrated with larger values related to national security and community building.

On the vertical dimension, the modern conception of water as property is loosely linked to territory because the holder, the nation-state, is defined on a territorial basis. Seen from this perspective, however, the state does not own water to ensure its own existence, but for the sake of its people, as a resource for social and economic development and growth. Under specific conditions, however, states adopt an outlook in which water is perceived as essential to the survival and continued autonomy of the sovereign, territorially defined state, especially when watercourses are seen as strategically important defensive boundaries or economic lifelines.

Not only territorially defined political units like sovereign states might take a more essential and "transmodern" stand on the issue of water than constructing it simply as a territorially defined property. For example, specific advocacy coalitions unrelated to territory (such as boaters, surfers, and fishermen) are sometimes bound together by certain core beliefs and understandings about the meanings of water. Water is essential for the existence and identity of social actors, whether they are acting as members of a territorially defined sovereign state or a nonterritorial community. The following paragraphs provide some examples of these transmodern meanings of water.

If water is perceived as something essential to national survival and the building of the nation-state, it becomes conceptualized as a security issue. In many places, water's intimate ties to security are particularly evident. Watercourses are frequently used to mark boundaries, and crossing them threatens the existence of a state. For example, the Rhine River has historically served as the dividing line between German and French nations and cultures. Control of the Rhine, therefore, has been a clear indicator of the balance of power between the two countries. When Bismark defeated the French in 1871, and the Germans occupied all of Alsace, Germany was able to develop the Rhine. This corresponded to the time when Germany could claim dominance in Europe. On the other hand, the defeat of Germany in two World Wars reestablished the Rhine as a boundary on German expansionism and reasserted French security.

Water is bound up with economic development, and in some parts of the world control over water is basic to a nation's potential to prosper. The history of conflict over water in the Middle East dates back thousands of years, and in spite of numerous attempts at conflict resolution in the region since 1919, water—or, more accurately, the lack of water—remains a major stumbling block to stability in the region (Butts 1999).

The semi-arid to arid nations of Israel, Jordan, Lebanon, and Syria are highly dependent on the water of the Jordan River basin (18,300 sq. km.) for agriculture, industrial production, and other economic development. Periodic droughts and rapid increases in population and areas under irrigation have exacerbated problems of water scarcity. Because 60% of Israel's supply of groundwater and 25% of its surface water supply originates in occupied territory, continued occupation of the West Bank assures Israel control over these critical water resources. The inequitable allocation of water among Israeli and Palestinian settlers in the West Bank, along with well drilling by Israeli settlers, have contributed to tensions in the region (Lowi 1999).

Efforts to promote peace in the Middle East through regional development of water resources have fallen short of their goals not only because of a failure to achieve political resolution of the Arab-Israeli conflict, but also because of conflicting water needs, international legal precedents, and perspectives on water rights. Mistrust, fears of dependency, and perceived threats to national sovereignty on the part of Jordan River basin nations have also made solutions difficult to achieve (Lowi 1999).

The end of the cold war has done nothing to diminish the importance of water as a security issue; indeed, it seems to have enhanced it. As the global population grows and demand for usable water presses closely on available supplies, the potential for conflict increases. Authors such as Jessica Tuchman Matthews (1991) and Robert Kaplan (1994) have painted grim portraits of the impacts of resource depletion and pollution on human societies and highlighted the potential for environmental issues to lead to violence and conflict. Environmental factors are now widely recognized as contributing to instability and conflict within and between nations (Renner 1997).

Even if water and other environmental resources were not scarce, other factors would provoke international confrontation. After the end of the cold war, the military/industrial complex began searching for new threats to legitimize its continued funding—and feuding—base. Potential conflicts over water are often constructed as excellent reasons to maintain military prowess. These constructions are vehemently contested by those who remind us of the hazards military conflict or preparedness hold for water resources within and between nations. The catastrophic damage to coastal water systems as a result of the Gulf War and the extensive surface and groundwater contamination around military reservations or nuclear weapons production sites in the United States and Russia offer graphic evidence of these hazards.

Just as water has a higher salience when it is perceived as a security issue rather than merely a property right, its importance also increases when it becomes a focal point for community building. Increasingly influential in intermestic politics, emerging transnational communities and coalitions embrace visions of water that disregard conventional political boundaries. As Joe DiMento illustrates in his chapter describing the evolution of the Black Sea Environmental Programme, restoring that sea's water quality has attracted interest far beyond its six adjacent littoral states. The Black Sea has brought to its shores tourists from all over Europe. This globally important water resource has also gained the attention of a number of international environmental nongovernmental groups. Funding for and participation in studies of the Black Sea have come from a variety of sources, some as distant as Denmark, Holland, Spain, and Japan. Indeed, the Programme's first coordinator was an academic from Britain.

Among entities who may use water as a focal point for building community are epistemic communities that serve as cognitive baggage handlers for knowledge, transmitting and diffusing new ideas among actors. These communities also provide a shared understanding of problem definitions and policy concepts across national lines (Haas 1992). Currently worldwide networks of scientists, other professionals, businesses, and interest groups coalesce around various notions of water resources. Shared interests and beliefs, not national identities, unite and distinguish these groups, and the relevant lines of interaction are those of open communication, not heavily fortified borders. The flows that matter in such groups are not cubic meters of water moving from one nation to another but information about water circulating between individuals and groups. Examples abound, including international societies of limnologists who study lakes, marine biologists who concern themselves with the characteristics of oceans as habitats, and international associations of lawyers who trace the evolution of treaties and conventions. On a more elemental level, water is integral to fishing communities. As Kate Sullivan reports in her chapter on the salmon wars, Canadian fisheries conducted a blockade of approximately 200 boats, evidence that they believed their way of life was being fundamentally threatened by the United States' refusal to negotiate agreements under or to abide by a treaty governing international salmon fisheries. Such action required group communication and solidarity.

One type of culturally bound community is frequently overlooked: groups that share a common interest in sports or leisure activities. Examples of such communities that might become involved in water policies are boaters, sport fishermen, divers, or white-water rafters. These groups transcend state boundaries, have a specific perspective on water, and make certain demands on quantity and quality of water and watercourses. Increasingly, they are important forces in political conflicts over the usage and distribution of water. Joachim Blatter provides an example in the resistance of transnational associations of sportboaters on Lake Constance to publicly owned shipping companies' planned use of motorized high-speed ferries (catamarans) on the lake. Since tourism is already the largest sector of the economy in all developed countries, these types of nonterritorial communities can no longer be ignored in attempts to understand conflicts about water.

Water can form the basis for building nonterritorial communities in nearly limitless ways. What makes these communities "essentialistic" in their relation to water is the centrality of water for their existence and their single-mindedness: A specific conception of water is the core of their shared belief systems. In contrast to broad-based political communities like the modern nation-state, these communities are single-issue coalitions with a very narrow range of goals and purposes. Consequently, game theory based on the assumption of strategic action cannot capture policy processes featuring political actors who possess fundamentalistic or essentialistic connections to water. Neither perceived threats to national security nor fundamental value conflicts allow for "rational" solutions like side payments or package deals when such actors are involved, as convincingly illustrated in Joachim Blatter's analysis of water governance politics at Lake Constance.

Products beyond Control: Natural Limits and Expanding Frontiers of Imagination

The modern conceptualization of water as a product of industrial and mechanical processes has been fundamentally called into question by newly emerging ecological realities. We, as humans, have been forced to confront mounting evidence that the carrying capacity of our planet, as well as our ability to control nature, is inherently limited. As the universe reveals itself to be far more chaotic and unpredictable than modern science once envisioned, the ecological paradigm provides fresh justifica-

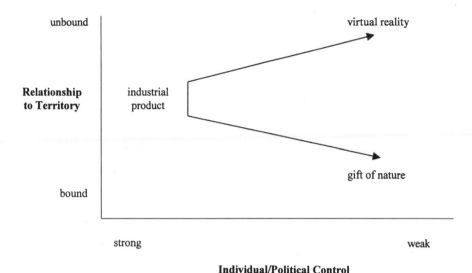

Figure 2.4
Beyond technical approaches: Diminishing control over water

tion of the ancient perception of water as a sacred gift of nature not to be manipulated by human efforts.

Ironically, amidst growing awareness of the natural limitations of human activity, we are simultaneously witnessing a transformation in conceptualizations of water toward a symbolic, virtual reality that is potentially more liberating from natural, material, or territorial conditions than the modern concept of technical production. The emerging virtual reality, characterized by a symbolized economy and social activities driven by the search for distinctive images, lifestyles, and tastes, will expand the meanings of water into dimensions formerly unimaginable. On the horizontal dimension of figure 2.4, the meanings of water range from water as a product of mechanical/industrial processes absolutely controlled by humans to water as something intractable to human manipulation. At the far right of the figure, water is liberated from the anthropocentric belief that humans can subdue the natural environment. Ecological and socially constructed complexities alike undermine the rationale for technical control and instrumental usage.

The overwhelming complexity and unpredictability of the postindustrial world mirrors that of the natural environment. In a landscape dominated by a symbolized economy and virtual realities, self-determination through control of the environment is illusory. Increasingly disconnected from material reality, assets determined by stock markets become products of the human mind: objects of expectations and speculations driven by multiple interdependencies and unpredictable waves and shocks. The rise of a specific water sport (like surfing) based on the symbolization of a distinct lifestyle is a dynamic and emergent phenomenon, socially constructed but beyond prediction and control.[1] For some people, surfing is a way of life that shapes friendships, determines dress, and introduces a common vocabulary of discourse. Further, surfing can be the basis of political mobilization against such threats as pollution of coastal waters.

The extent of water's association with territory is illustrated on the vertical dimension of figure 2.4. The modern conception of water as a material product is only loosely coupled to territory. Though it is no longer bound to its place of origin, the materiality of water nonetheless limits its portability. As a gift of nature, rainfall and rivers are quite localized. From this perspective, water is inseparable from place, and attempts toward transport of water destroy the essence of both the areas of receipt and origin. Simply put, water belongs in the ecological context where nature placed it, and humans should adapt accordingly.

Water as a gift of nature is a theme repeated a number of times in the case studies in this book. Premodern examples of this theme are illustrated in two of the chapters following this one. To the Cocopa, who made their living in the Colorado River delta prior to European settlement, water was an intimate part of life. María Rosa García-Acevedo describes this 2,000-year-old community of fisheries and flood recession farmers in chapter 3. As David Hughes reports in chapter 10, the native residents of Zimbabwe also saw water as a natural feature outside human control. Linear water, in this case, provided natural demarcation on which the tribal peoples' notions of territory were based.

The postmodern perspective of water as a natural element beyond human control is evidenced in other chapters of the book. In chapter 6, one of the discourses Kate Sullivan discovers to be important in the governance of wild North American salmon resources relates to complex biological models mapping the geographical boundaries and sustainable populations of various salmon runs. The message from the life sciences is clear: Fisheries have natural limits. In chapter 5, Suzanne Levesque depicts water as an integral part of the natural landscapes that are conceptualized as habitat and wildlife movement corridors by the Yellowstone to Yukon Conservation Initiative. Land fragmentation threatens the kind of holistic land management the movement views as essential for the preservation of biodiversity. The inability of large mammals, like bears, to survive under current conditions is a signal of deeper ecosystemic imbalance. From this group's perspective, human intervention into natural processes may have dire consequences, not only for water but for the entire "web of life."

Water has other meanings, not firmly bonded to territory, over which human control is clearly limited (see figure 2.4). These meanings are undergoing cutting-edge changes, the implications of which are as yet not entirely clear in relation to transboundary water resources. To get a sense of what is to come, we want to trace the emerging transformation of water from a material reality to a symbolized and virtual reality along three lines:

1. the transmogrification of water from an input factor for agroindustrial processes to a lifestyle product (like the French table waters Perrier or Évian);
2. the transplantation of "natural experiences" from natural places to adventure/theme parks and, most recently, cyberspace; and
3. the move from national water projects to global water banks.

Before addressing the changing character of the instruments and actors involved in water policy, it is necessary to describe the transformation of water as a product. Imbued with symbolic values, water has been reimagined as an indicator of taste and lifestyle. This shifting image has many consequences for water policy. For example, a much higher environmental sensitivity characterizes the production of table water like Perrier or Évian than that surrounding the use of water in an industrial process. Intriguingly, these companies, blessed with fashionable brand names that facilitate access to public discourse, might be more powerful than others that use much larger amounts of water. This phenomenon is well exemplified by the contemporary sale of bottled table water from Lake Constance, the subject of chapter 4. Thirty years after the German news media reported the lake's near collapse due to environmental pollution, Lake Constance's largest water company now complements and differentiates its lake water distribution and marketing strategy. Although the company continues to transport billions of cubic meters of water to its customers through its pipelines, it now offers the same water in bottles—for a much higher price. Bottling the water of Lake Constance is a profitable enterprise that owes its success to the imagery of Lake Constance as the sunny "Riviera" of Germany, complemented by the optimistic tale of the environmental restoration of the lake.

Another example of water as an artifact of human imagination is the artificial and virtual water created by entertainment and commercial industries. A first step in the process of transforming water into an artifact of imagination has been the building of adventure and theme parks displacing the human experience of water from its natural environment, making it more accessible and comfortable. A second and even more profound displacement of water from its embeddedness in nature has been the creation of virtual water through modern telecommunication and multimedia. The sounds of fountains, brooks, and waves are captured from nature or synthesized in sound studios and marketed to customers as "natural music" with relaxing properties. Videotapes of cascades and waterfalls are standard backdrops for selling beer and soft drinks because they conjure in the viewer's imagination the image of something refreshing and thirst quenching. The most recent displacement has been the global accessibility of these virtual representations of water in the cyberspace of the Internet.

These developments raise a host of intriguing questions. How far is it possible to redirect the demand for water as a medium for sport, entertainment, and adventure from natural watercourses to artificial and virtual waterplaces? What does this mean for water policies? In many cases, alliances between tourism and environmentalism have been very important in battles for the protection of watercourses. Unlike physical water, there is no scarcity of virtual water, and no danger of virtual water pollution.

Water banks, a concept many environmentalists and economists advocate, would allow water not used at a particular time, or water imported from another area, to be reserved or "banked" in underground aquifers for future use. Such reserves or investments are currently possible only in fiction, since as a physical entity water is a flow resource that is constantly in motion, unless captured in a reservoir. For water banks to work, lenders and borrowers would have to be convinced that water could be transformed into currency or contract. The actual introduction of water banks would fundamentally transplant the locus of decision making in water projects from the political arena to the market sphere.

One of the most important transformations that will alter the meanings of water is the metamorphosis of municipal water companies into multinational corporations, a process that is most advanced in France. Vivendi (formerly Compagnie Général des Eaux) and Lyonnaise des Eaux are, at this time, two of the largest publicly traded French companies. Their privatization has been accompanied by decisive steps toward globalization and symbolization: Lyonaise des Eaux, for example, is now providing water for Casablanca, Indianapolis, and Rostock (Germany) and is starting to do so in Potsdam, Manila, Budapest, Jakarta, and Ho Chi Minh City. These companies, which originally provided water services only in their home cities, are now rapidly diversifying, especially into the field of telecommunication (*Frankfurter Allgemeine Zeitung*, August 22, 1998). Consumers will soon be able to purchase "real" water and its virtual counterpart from the same entity. In the future, these global players and their perceptions of water will greatly influence (transboundary) water policy.

Water banks and water companies traded on the stock market can be envisioned as quite modern phenomena that will increase the possibilities of rational allocation and efficient use of water. What marks them as postmodern phenomena are not just the dual realities of symbolization and dematerialization but also their vulnerability to "hyperrationality." Like many elements of financial markets, these instruments promise to improve allocative efficiency while reducing uncertainty. But in the marketplace, these attempts at rationalization are often transformed into instruments of speculation. For a water company, the value of its stock (or its market capitalization) and its corresponding capability to invest in large projects is determined more by the capriciousness of the market than the "real" situations at the places proposed investments will occur.

In recognition that our world is largely beyond control, research on transboundary water policy must take into account the insight that political action is very often a reaction or adaptation to natural and social crises rather than strategic action guided by instrumental rationality. Agent-based modeling, informed by the assumption that agents use adaptive rather than optimizing strategies (Axelrod 1997), is one of the most promising research approaches incorporating this insight. Explaining the evolution of cross-border policies and institution building by looking at spillovers of connected political arenas to the cross-border arena (Blatter 1997 and in this volume) is another attempt to take these transmodern realities seriously. Furthermore, acknowledging the growing importance of information and symbols leads to stronger emphasis on discourse and communication in the analysis of water policies (Sullivan, Doughman, and Levesque in this volume).

From Standardized Commodities toward Connected Specific Goods

Modern thought, with its emphasis on precepts of economics and rationality, has tried to liberate water from its ancient territorial bonds in order to treat it as a tradable standardized commodity. As such, water has partly escaped its ties to territory and its limitation to use only in its natural form. Its numerical price in the marketplace provides the universal mechanism for transferring water from one place to another, from one usage to another, and from one time to another. Today, increasing awareness of old and new limits for water exchanges across places, times, and purposes has seriously undermined the modern belief in universalistic usage of goods through standardization and homogenization. Figure 2.5 illustrates in the horizontal dimension departures from the modern conceptualization of water as a commensurable, fungible, and exchangeable commodity. Water can become incommensurate and a very "specific good" (Williamson 1975)[2] in various ways.

From an ecological perspective, water is, through its connection to its natural environment, in many ways a specific good. Water carries with it the imprint of its place of origin, including various types of microbial life and dissolved solids, temperature, corrosiveness, and taste. The dangers of altering the environment at the places of origin and destination place serious limitations on the commensurability of water. Water is also a specific good according to the cultural perspective, because it is deeply embedded in communal life. The noncommodity meaning of water may

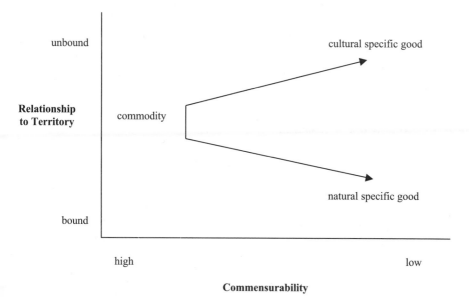

Figure 2.5
Beyond economic approaches: The uniqueness of water

depend very much on shared beliefs transported by symbols, religion, and myths.

Both kinds of embeddedness—the connections of water to both the natural and the social environments—oblige researchers, in many cases, to view water not as a commodity but as a specific good. The vertical dimension of figure 2.5 depicts the varying relationships of water to territory. Whereas ties to the natural environment obviously have a territorial basis, cultural ties are no longer strictly bound to territory. Though countless examples show that specific cultural meanings of water are rooted in traditional elements like religion, myth, and rituals, the processes of glocalization provide opportunities and incentives for potentially far-flung communities to create new symbolic bonds to water unbound from territory.

As noted above, water carries with it the imprint of its place of origin and the human processes to which it has been subjected. Experience with the Glen Canyon Dam on the Colorado River has taught us that water held behind dams may lose many of its natural properties. Alterations in the natural characteristics of water, may result in significant changes in places. Throughout the Grand Canyon, the banks of the Colorado River have been robbed of their sandbars and sidepools because the water released from the dam carries a different sediment load than naturally flowing water. The colder water released from the bottom of the dam favors the growth of trout, displacing native species such as the squaw-fish and the humpbacked chub.

This book offers instructive examples of transboundary waters affected by human interventions. For instance, the extensive damming and diversion of the Colorado River, resulting in higher evaporation rates and more saline return flows to the river from irrigated agriculture, so modified the salt levels in the river that its waters became virtually unusable in Mexico. As María García-Acevedo explains in chapter 3, Mexico and the United States consequently engaged in tense negotiations about the terms of the 1944 treaty between the United States and Mexico that allocated an agreed-upon quantity of water to Mexico. The United States maintained that water was water, that Mexico was receiving its proper allocation, and that the 1944 treaty stated nothing about the quality of the water Mexico received. Mexico, however, maintained that the waters of the Colorado River had historic characteristics, and when the treaty referred to water, it meant the kind of usable water Mexico had come to depend on.

Historical, anthropological, and contextual case study analysis suggests that water has a communal value that transcends its value as a commodity (Brown and Ingram 1987). This communal value is often described as being tied to place. For the indigenous Apaches, water is often imbedded in place names on reservations, such as "water flows inward under the cottonwood tree." These kinds of names evoke not so much a description of the natural setting as the oral tradition of stories handed down from ancestors, each of which carries a moral lesson of tribal significance (Basso 1996). Another example of the communal value of water is provided by the Ganges, a fundamental element in Hindu culture, which is seen as a goddess who purifies all those faithful who immerse themselves in her. Some 60,000 devotees ritually bathe in the Ganges each day in the city of Varanasi, lighting fires to Lord Shiva, who is said to have caught the river in the tangled locks of his hair as he descended to earth from heaven (Stille 1998).

Lake Tahoe, Niagara Falls, Old Faithful, the fountains at Tivoli Gardens, the canals of Venice, the Cote d'Azur, Norwegian fjords, Victoria Falls—all have value far beyond whatever expense might be incurred by spectators to experience such treasures. The nearly universal acknowledgment and appreciation of water as a central element in nature and a source of natural beauty explains the dedication of public preserves, national parks, and wild and scenic rivers for the enjoyment of humans. These places are sacrosanct not primarily because of their ecosystemic value but because of their socially constructed aesthetic value. Therefore, they have to be defined as cultural specific goods or as "cultural monuments." Independent of the question of how "natural" the water at these places, which range from unaltered waterfalls to artificial fountains, actually is, they are seen as public goods connected to specific places. The value of water and watercourses that have attained status as cultural monuments is beyond economic calculation. Even if cost-benefit analysis revealed that using water for production of electricity or for shipping would prove more profitable than tourism, the uniqueness of such cultural monuments delegitimates such profane calculations.

Movements to protect cultural monuments could be seen as new kinds of "imagined communities." Imagined communities rely on ideologies or belief systems rooted in place when, in fact, they are not territorially bound; they transcend the boundaries of geographic regions and nation-states. Pam Doughman's discussion in chapter 7 of sustainability criteria being applied by international development banks and new transboundary environmental governmental organizations is relevant here. The notion that what is done with water under international auspices must meet a culturally constructed standard is a new and potentially tranformatory concept. These concepts are also illustrated in Suzanne Levesque's discussion in chapter 5 about the Yellowstone to Yukon Conservation Initiative, a non–territorially bound environmental network that transcends the boundary between Canada and the United States. The network draws supporters from widely disparate regions throughout both nations who advocate the preservation of space within a specific geographic region.

Because of the role that water plays in constructing cultural identity, research on transboundary water policy has to take into account the idea that water as a cultural specific good is quite often beyond the logics of commensurability and exchange. The processes of glocalization are making profound changes in water policy dynamics as globalized advocacy coalitions and the new significance of local and regional specificities compel acknowledgment of water as a cultural specific good.

Implications for Future Water Research and Management

This chapter has chronicled the expanded range and diversity of meanings of water in the contemporary political and academic realms. We have argued that the definitions of water imposed by the legal, technical, and economic approaches characteristic of modern instrumental rationality are no longer solely authoritative. The central focus these approaches once provided will increasingly be forced to give way to a broader array of approaches. The confines of modern approaches will be debordered in future attempts to analyze water politics in a glocalized world.

The case studies in the chapters that follow demonstrate that premodern and modern meanings of water continue to be relevant in present times even as postmodern meanings are becoming more significant. Water is frequently viewed in the subsequent chapters as a focal point for community building. Water is also closely associated with culture. Virtual and symbolic water weighs more heavily in case study analysis than water as property, product, or commodity. We must understand the meanings of water held by the social actors involved in water conflicts, as opposed to simply imposing a modern Western definition of the problem. This leads us to propose more qualitative research approaches and methods like historical analysis, network analysis, and discourse analysis as described in chapter 1. At least as an initial step of investigation, these approaches should be given predominance over narrow rationalist approaches with constrained definitions of actor identities and simple utilitarian preferences with respect to water.

We are convinced that this kind of reflective scholarship will become even more necessary in the future when arguments over the meanings of water will assume a central place, frequently intensifying into conflict, misunderstanding, and injustice. The actors able to impose their pre-ferred definitions of water upon discussion and debate will have the upper hand in determining the outcomes.

Nowhere is the contest over divergent meanings of water likely to be more interesting and informative than in transboundary situations. As chapter 1 illustrates, the decline of the nation-state as the gatekeeper between the domestic and the international realm has led to a situation in which numbers of new and emerging actors are crowding the field of transboundary relations. Some nation-states and civil actors may still ascribe to conventional notions of water as property, product, and commodity, but they meet strong resistance in attempting to forge imple-mentable agreements. Other actors, such as local governments, commu-nity groups, and transnational environmental networks, often reject the premises of these nation-states and civil actors and envision what is at stake in wholly incompatible and noncommensurable ways. Compro-mise under such circumstances becomes virtually impossible as con-testants talk past one another. Little progress toward understanding and potential resolution of water problems is now possible without first clarifying and accepting as legitimate the widely varying meanings and values associated with water.

Having argued against a narrow conception of water throughout this chapter, we will nevertheless conclude with a statement that may initially seem to be contradictory. We do believe that, in many circumstances, the modern "disembedded" meanings of water are the most helpful for find-ing solutions and compromises across various "boundaries"! This is be-cause disembedded meanings of water like "product," "property," and "commodity" are not essentialistic as meanings like "security," "iden-tity" and "specific goods" are. In "modern" conflicts a solution can be found by bargaining and deduction from general (scientific or legal) principles and rules. In contrast, framing water conflicts as security issues may well lead to pure power battles. Water as a focal point for transnational community building may well broaden the already deep, normative cleavages within societies. Consequently, the most productive strategy for analysts and actors may be to attempt to translate or shift nonmodern to modern meanings. However, such a strategy is often impossible and can be accomplished only if the existing meanings of water and their roots are completely understood through the approaches we suggest here. In many instances, one has to acknowledge that reinterpreting pre- and postmodern meanings as modern meanings is not feasible. Consequently, modern solutions (such as side payments or legal standards) are unrealistic and misguided.

We contend that scholarship should contribute to the management of transboundary water resources by:

1. helping to build bridges between various meanings and understandings. This can be accomplished primarily by proposing instruments like mediation and purely procedural legal regimes that, instead of predetermining any substantial meaning, are specifically framed to empower underprivileged voices (the critical, reflective modern perspective that still considers the possibility of positive solutions through collective action).

2. enhancing the legitimacy of noninstrumental uses of water and defending specific natural and cultural meanings of water. This can be accomplished by explaining specific natural and cultural meanings of water and defending them against modern attempts to transform and manipulate such meanings (the critical, transmodern perspective opposing the myth of the best possible solution).

Which perspective and approach each scholar may choose is a matter of personal values. However, contrary to modern approaches, critical scholarship does not hide its normative positions behind a veil of objectivity.

Notes

1. Control has two implications: One can steer the direction of the process outcome, and one's input is responsible to a large degree for the outcome. If one's activities are decisive for the outcome but one cannot steer the direction of the outcome, there is manipulation but no control. If one contributes to the process with one's activities in a certain direction but one's input is almost negligible to the outcome, one participates but does not control.

2. We borrow the label "specific good" from Williamson (1975) even though we go beyond his narrow technical and economic understanding of specific goods. Specific goods are characterized by the whole being more than the sum of the

parts and by high transaction costs. This corresponds to our emphasis on the connectedness of water to its natural and social environment and to the difficulty and sometimes impossibility of finding a price for such a good.

References

Axelrod, Robert. 1997. *The Complexity of Cooperation.* Princeton: Princeton University Press.

Basso, Keith H. 1996. *Wisdom Sits in Places: Landscape and Language among the Western Apache.* Albuquerque: University of New Mexico Press.

Blatter, Joachim. 1997. "Explaining Cross-Border Cooperation—A Border-Focused and a Border-External Approach." *Journal of Borderland Studies* 12, no. 4: 151–174.

Brown, F. Lee, and Helen M. Ingram. 1987. *Water and Poverty in the Southwest.* Tucson: University of Arizona Press.

Butts, Kent Hughes. 1999. "The Case for DOD Involvement in Environmental Security." In Daniel H. Deudney and Richard A. Matthew (eds.) *Contested Grounds: Security and Conflict in the New Environmental Politics.* Albany: SUNY Press, pp. 109–126.

Haas, Peter M. 1992. "Introduction: Epistemic Communities and International Policy Coordination." *International Organisation* 46, no. 1 (Winter): 1–35.

Kaplan, Robert. 1994. "The Coming Anarchy." *Atlantic Monthly* 273 (February): 44–76.

Lowi, Miriam R. 1999. "Transboundary Resource Disputes and Their Resolution." In Daniel H. Deudney and Richard A. Matthew (eds.), *Contested Grounds: Security and Conflict in the New Environmental Politics.* Albany: SUNY Press, pp. 223–245.

Matthews, Jessica Tuchman. 1991. "Redefining Security." *Foreign Affairs* 68 (Spring): 162–177.

Renner, Michael. 1997. "Transforming Security." In Lester Brown et al. (eds.), *State of the World 1997.* New York: W.W. Norton.

Stille, Alexander. 1998. "The Ganges' Next Life." *New Yorker* (January 19): 58–67.

Williamson, Oliver E. 1975. *Markets and Hierarchies, Analysis and Antitrust Implications: A Study in the Economics of Internal Organization.* New York: Free Press.

II

Case Studies

3

The Confluence of Water, Patterns of Settlement, and Constructions of the Border in the Imperial and Mexicali Valleys (1900–1999)

María Rosa García-Acevedo

Images of water continually float to the surface of popular discourse about the U.S.-Mexico border. Waves or tides of illegal immigrants are said to be streaming or flooding across the leaky, porous U.S.-Mexico border; some say the border is as useless as a breached dam in holding back the deluge of people pressing against it. "Drowning in immigrants" and "immigrants drowning" are meant figuratively as well as literally. The deprecatory term "wetbacks" derives from the way in which many undocumented persons cross the Río Bravo/Rio Grande into Texas, a deadly, dangerous journey during flood season. The repeated and sustained use of water imagery in discourse on the U.S.-Mexico border springs from a deep understanding of how water has affected the settlement patterns and way of life of people in a border region and the character of the border societies that have resulted.

This chapter engages the in-depth, historical case study method to examine the changing meaning of water between 1900 and 1999. This method of analysis allows for context to take the central place in understanding transboundary water resources on the U.S.-Mexico border. The chapter contends that the meanings ascribed to water are directly linked with migratory patterns of people. The currents of water flows have actually carried people from one place to another and pushed them into particular kinds of employment and living conditions. Moreover, the political boundary separating the two countries has served as a "headgate," opening and closing to control the flow of people and commerce according to the particular meaning(s) and management of water prevalent at different points in time.

The inextricable connection of water to people and the border is nowhere in greater evidence than in the area of the case study, the Imperial and Mexicali Valleys, which are proximate to the lower delta of the Colorado River. What writer Otis B. Tout observed about the Imperial Valley in 1931 is an inherent truth in the binational region: "[It is] not only the very lives of the people of the Imperial Valley [that] depend upon the safe and secure flow of water through the irrigation system, but also their social welfare, their culture, and their ability to pursue and attain happiness and success" (Henderson 1968, 131).

In chapter 2, Joachim Blatter, Helen Ingram, and Suzanne Levesque introduced the notion that the meanings of water are expanding in the contemporary world, with modern meanings being supplemented and at times displaced by new meanings. When the framework developed in chapter 2 is applied to the Imperial and the Mexicali Valleys, two important concepts surface.

First, at any point in time, local groups attempting to construct new meaning(s) for water may find that they have less autonomy and freedom in this area than they suppose. Their very existence is a result of actions and reactions to previously dominant meanings. A close link exists between water, equity, and social engineering.

Second, an examination of the Imperial and Mexicali Valleys suggests that there is often a discontinuity or lag time in movement of redefinitions of water from one side of the border to the other. This chapter explores the definition and nature of the international dividing line and contends that meaning changes initiated in one place spread to the other side, where they may be implemented at a different pace and benefit various groups and interests at different times.

This chapter identifies the winners and losers among contending interest groups in the water policy realm, but more important, it illustrates the preconditions that led to changes in constructions of water and the consequences of each construction for flows of people in the Imperial and Mexicali Valleys. Prevailing definitions of water spring from previously channeled sources. The lessons of the history of the Imperial and the Mexicali Valleys suggest that it is an illusion to suppose that people control water. More accurately, water touches everything, and to fail to understand how and why is to be left completely at its mercy.

Water as a Gift of Nature

Like other rivers in other areas of the world, the Colorado River was "the most important natural factor influencing native cultures in the delta" (Castetter and Bell 1951, 4). Water was an intimate part of the life of the Native populations, including the 2,000-year-old Cocopa community of fishers and flood recession farmers, and was not separate from the land, wildlife, or any aspect of the natural and human world, which they saw as a unified whole (Morrison et al. 1996, xi; Alvarez de Williams 1974, 100–104). The name "Cocopa" in the native language means precisely "those who live in the river" (Alvarez de Williams 1974, 68).

The extensive seasonal flooding of the lower Colorado River deeply affected the lifestyle of the Cocopa. Permanent fields could not be maintained; thus, periodic moves from place to place to engage in agricultural activities were a common practice. Figure 3.1 shows where Cocopa bands were located in the nineteenth century.

The displacement of the Cocopa began with the invasion of outside settlers that started in the early nineteenth century. After 1848, the territory of the valleys was increasingly explored by both Americans and Mexicans, who would have found the hot, arid area unsuitable for settlement were it not for the potential of the Colorado River. Instead of following the dictates of the river, newcomers, both individuals and companies, determined to harness the river for human uses as a course for transportation and later for irrigation. They ignored the Cocopa's historical rights as the first users of the river and found the tribe of interest only as a source of cheap labor (Alvarez de Williams 1975, 45–133).

Through changes in the use of land and water, the newcomers annihilated the mezquitales and other plants that provided food for the Cocopa. Their use of the waters of the Colorado River for irrigation limited the overflow of water, thereby interrupting the traditional agricultural and fishery activities of the Cocopa (Castetter and Bell 1951, 8; Morrison et al. 1996, xi; Alvarez de Williams 1974, 101–103). As a result of this disruption, many of the Cocopa slowly ventured into the outside world. Many cut wood that later was sold to the Colorado Steamship Company, whose boats navigated the Colorado River (Fradkin 1981, 294).

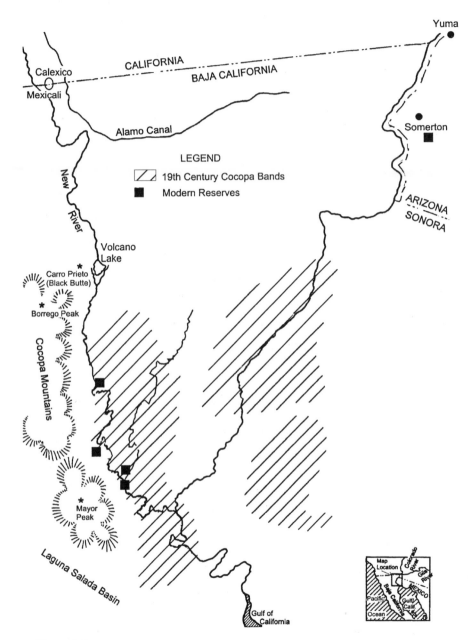

Figure 3.1
Nineteenth-century and modern Cocopa reserves

After 1877, when the extension of the railroad to Yuma bankrupted the steamboat companies, some Cocopa looked for jobs in agriculture.

The most decisive blow to native lifestyles was dealt in 1905 by a mistake in canal headgate engineering by the California Development Company that resulted in the diversion of the entire flow of the Colorado River—which flooded into the Imperial and Mexicali valleys for the years between 1905 and 1907—and completely reconfigured their homeland. As a result, some groups of Cocopa headed to the Yuma Valley to join those who had emigrated from the Cocopa homeland earlier. In 1910 and 1917, the U.S. government allocated land for the Cocopa reservation on the outskirts of Somerton, Arizona (Castetter and Bell 1951, 56; Kelly 1977, 13; Alvarez de Williams 1974, 68). Other members of the Cocopa settled south of Mexicali in *rancherías*. They toiled in the fields of the valley as farm workers (Alvarez de Williams 1975, 133; Castetter and Bell 1951, 53–55; Estrada Barrera 1978, 63; Estrella 1982, 1).

From the mid nineteenth century to the late 1930s, however, the Cocopa moved freely across the U.S.-Mexico border without encountering any obstacles. Even the Gadsden Purchase (1853), which imposed formal borders on the Cocopa's traditional homeland, did not bring immediate changes for the community (Kelly 1977, 13; Alvarez de Williams 1974, 36). Norwegian geographer-explorer Carl Lumholtz documented in 1910 how those Cocopa who were part of the community in the Yuma Valley returned to the Mexicali Valley every year "when the River rises to plant crops" (Alvarez de Williams 1975, 133).

By the end of the 1930s, however, the population movements of the Cocopa were increasingly limited. Responding to the complaints of the Yuma County Board of Supervisors, the Immigration and Naturalization Service (INS) initiated constant raids in Somerton. The Board argued, in response to the increased number of Cocopa who visited or moved permanently to Somerton when they were unable to secure a job in Mexico after agrarian reform took place there in the late 1930s, that the "Indians" in the area were a "welfare burden." As a consequence, Mexican-born Cocopa encountered increasing limitations in crossing the border, even temporarily. Thus divided, the Cocopa communities on both sides of the border began to grow apart.

Finding it completely impossible to persist in pursuing a premodern definition of water as a gift of nature, the Cocopa have tried with very little success to gain a foothold in the region's political system. For example, the Cocopa living on a U.S. reservation in Somerton have faced water problems: The allocation of land and water resources for the reservation (242, 709 acres) was insufficient to provide an adequate support base for those who lived there (Estrada Barrera 1978, 63; Kelly 1977, 13).

Persistent effort to lay claim to sufficient water for development by the Cocopa, together with similar efforts by other tribes, eventually yielded some success in the United States. In 1963, the Supreme Court in *Arizona v. California* quantified the water entitlement of five tribes on the lower Colorado River. The Court's decree allocated water to each of the five reservations according to a certain standard, with priority dates corresponding to the date that various lands were included as part of the reservations (Checcio and Colby 1993, 45). Of course the potential use of such water in modern irrigation is very different from and incompatible with the premodern meaning of water as a gift of nature.

Currently the Cocopa Reservation is divided into three parcels, known as East Cocopa, West Cocopa, and North Cocopa. About 2,400 acres of irrigated land exist on the reservation. Most of the land, however, has been leased to nontribal farmers. With the acquisition of 4,200 acres in 1985, the Cocopa started new businesses, such as convenience stores and gas stations (Cocopa Indian Tribe Website).

Beginning in the 1930s, Mexico's policies also discouraged the Cocopa's transboundary links. Official discourse advocated the integration of the Native populations into the "Mexican nation" as the only way to overcome their "poverty." With the blessing of Mexican President Lázaro Cárdenas, an *ejido* (collective farm) for the Cocopa was created in the outskirts of Mexicali. However, the mismanagement of the project caused the effort to fail (Kelly 1977, 13). In addition, during the same era, Native populations—including the Cocopa—became practically invisible in Mexico. With the exception of noting speakers of the Cocopa's native language, which a small minority of Cocopa have managed miraculously to preserve, the Mexican census did not collect information on indicators of Cocopa ethnicity after 1930.

Similar to what has happened in the United States, in Mexico the Cocopa, along with other native populations, have been struggling with limited success to recover their historical rights to land and water. During the early 1970s, the Cocopa obtained certain guarantees from the Mexican government regarding the possession of land in Baja California, the *Decreto de Tierras Indias* (Decree on Indian Land). Currently, the main Cocopa settlement in Mexico is located in El Mayor, "a scattering of plywood and cement block shacks along a few dirt streets" (Rice 1997, 3).

Changes in the meaning of water away from "gift of nature" have affected the Cocopa perhaps more than any other peoples. Not only have they been displaced from historic areas of settlement, their livelihoods changed, and their ability to move around their historic homeland blocked, but their integrity as an ethnic group has also been jeopardized. As Anita Alvarez de Williams has observed, without "their river," the Cocopa vision of water as integral to community is impossible to sustain or recreate (Alvarez de Williams 1997, 354; 1974, 71).

Water as a Product in the Valleys: Early Developments

Private entrepreneurs who looked at the vast arid lands in the Imperial and Mexicali Valleys in the first decade of the twentieth century and the adjacent untamed Colorado River saw opportunities to create wealth on lands considered useless up to that point. Under the assumption that nature is malleable, they envisioned great water works that could bring water to the land. These capitalists conceived water as a product that could be manipulated through a system of dams and irrigation canals and sold at a profit to settlers. The challenge was to raise money for construction, employ the necessary engineering expertise, and profitably market to settlers a bundled package of land and access to water. Entrepreneurs with vision and ambition operated in both the United States and Mexico. Because capital was more readily available in the United States, U.S. investors dominated water and land development efforts in the area.

In the United States, speculative, privately financed irrigation ventures were cropping up all over the West after 1880. In the Imperial Valley, the California Development Company was created in 1896 with the purpose of irrigating the land west of the Colorado River. The company capitalized on the technical knowledge that freelance engineers and other professionals provided.

The California Development Company's main purpose was to build irrigation projects that would permit settlers on the newly irrigated lands. Meanwhile, the Imperial Land Company, a sister company founded in 1900, developed a policy oriented to attract prospective colonizers from all over the United States to the Imperial Valley. These colonizers were made stockholders in the company (Estrella 1982, 11–16). The first wave of immigrants came from the Salt River Valley in Arizona. The foundation of Calexico in 1901 took place as soon as water was available (Callahan 1967, 462; Gottlieb 1988, 95; Worster 1985, 196). Figure 3.2 illustrates the establishment of other agricultural communities, such as Brawley, El Centro, and Holtville, at various locations along the main irrigation canals.

In Mexico, the flow of people to Mexicali started in 1902, when the canals were opened (Aguirre Bernal 1966, 51). It is symbolic of the close links between Mexicali and Calexico since the time of the cities' founding that U.S. engineers were in charge of designing the city using similar blueprints to those used in Calexico—founded the previous year.

Floods in 1905–1907 destroyed Mexicali and halted the flow of population to the area. In addition, as a response to the financial crisis in the California Development Company as a result of flood damages, other U.S. interests were drawn to the Mexicali Valley. The Colorado River Land Company, founded by a group of U.S. entrepreneurs, including owners of the *Los Angeles Times*, bought up most of the land holdings from the California Development Company and became the most important player in the valley.

The Colorado River Land Company's policies greatly influenced immigration and settlement patterns in the Mexicali Valley. Newcomers to the valley were particularly affected by the company's reluctance to allow lessees or farmworkers to establish any type of permanent settlements on its land. As figure 3.2 illustrates, the city of Mexicali became the only urban area in the valley. This pattern of population concentration continues today.

Figure 3.2
Contrasts in density of human settlements in Mexicali and Imperial Valleys

As soon as it purchased the land, the Colorado River Land Company made its presence as a landlord known to residents on its newly acquired property. The company sent a message to the Mexican residents who had settled close to the delta of the New River (approximately 400 people), announcing that they had to move out or pay an exorbitant rent. The company wished to attract workers to the Mexicali valley and actively promoted the immigration of Asians (from Canton and Fujian, China; Punjabi, India; and various islands in Japan). The first Asians arrived as early as 1902. To justify its reluctance to hire Mexican laborers, the company portrayed Asians as highly reliable people. Underlying the construction of such a positive image was the conviction on the part of the Colorado Company that their status as aliens made them docile and prevented any undesirable social uprising on their part.

The construction of water as a product led the entrepreneurial interests in the United States and Mexico to espouse an open, permeable border between the two countries. Because of the geography of the land in the area (that is, its extensive sand dunes), construction of a canal on the U.S. side of the border to bring water to the Imperial Valley from the Colorado River was seen as technically difficult and prohibitively expensive (see figure 3.2). Thus, at the turn of the century the U.S.-based California Development Company enthusiastically supported the idea of using an overflow channel of the Colorado River in Mexico, the Alamo Canal, to irrigate property in the (U.S.) Imperial Valley (Fradkin 1981, 268; Henderson 1968, 17). A complex web of transnational business, which included the California Development Company, its counterpart in Mexico the Sociedad de Irrigacion y Terrenos de Baja California, and later the Colorado River Land Company, facilitated such arrangements. Through these arrangements designed by private companies, Mexico had access to one half of the Colorado River from the turn of the century until 1944 (when the governments of the United States and Mexico signed a bilateral treaty on water). During these years, the Colorado River Land Company, the main landholder in the Mexicali Valley, which also possessed land in the Imperial Valley, clearly profited from the water that Mexico received (Aguirre Bernal 1995, 341–342; Estrella 1982, 7–8; Gottlieb and FitzSimmons 1991, 76).

Even though the first Customs' checkpoint between Calexico and Mexicali was established in 1903, an open border, in fact, existed between the United States and Mexico. The Mexican consulate in San Diego reported the constant, informal crossing of the border. Neither people nor water encountered any barriers in such crossings (Walther 1996, 49).

With few exceptions, U.S. and Mexican authorities were only shadow agents in the early arrangements related to land and water that took place in the Imperial and the Mexicali Valleys. Although there was some discussion in the U.S. Congress about maintaining the "navigability" of the Colorado River, as opposed to allowing private interest to use its waters for commercial agriculture, the U.S. government did not offer strong resistance to the Alamo Canal project (Aguirre Bernal 1995, 341).

The hegemony of the U.S. private interests in the Mexicali Valley also met with the tacit approval of Mexico's government. Further, the Mexican government pursued policies in its dealings with the Colorado River Land Company that oscillated between open support and friendly negotiations. The only exception (which is discussed later) took place during President Lázaro Cárdenas's regime in the late 1930s.

While the entrepreneurial water companies masterminded the flow of water and people in the valleys, their actions also initiated the conditions for change through social mobilization of those attracted to the area. In the United States, landholders in the Imperial Valley—who founded the Imperial Valley Water Users' Association—struggled against the California Development Company for the control of water after 1902. Their complaints focused on the high price of water and the problems with its inadequate supply. The Users' Association lobbied the Bureau of Reclamation to initiate public water projects. The 1905–1907 floods (discussed below) were the catalyst that led to governmental intervention. The association argued that the disaster was evidence that a water canal through the U.S. rather than Mexican territory was essential to secure sufficient quantities of water to ensure crop production.

A number of grassroots movements also sprang up in Northern Mexico during the first decades of the twentieth century against the enclave of the Colorado River Land Company. During the Mexican Revolution

(1915–1920), Colonel Esteban Cantú, a prominent military man and *jefe político* (a powerful political personage), supported agrarian reform in response to the plight of Mexican farmworkers. These workers had emigrated from various areas of Mexico or returned from the United States to Mexicali. Cantú imposed certain limits on the power of the Colorado River Land Company, but he received no support from the Mexican central government for reducing the company's influence in Mexicali. Later, other *agrarista* movements emerged in the Mexicali Valley with no greater success, except for token concessions of land in response to their demands.

Construction of water as a product shaped immigration patterns in the Imperial and the Mexicali Valleys. Large numbers of persons, including Asians, flowed to the area as engineering works delivered water for agriculture. People settled where entrepreneurial interests wanted them to settle, spread across a number of agricultural communities in the Imperial Valley, and concentrated in the city of Mexicali in Mexico. Because it served the needs of water production processes during this era, the border was treated as quite permeable. U.S. investment flowed across the border unhindered, and irrigation water bound for U.S. users flowed without restriction through Mexican territory.

Water as a Security Issue

Water may be constructed as critical to national survival under certain circumstances and draw the attention of policymakers at the highest level of central government, as was suggested in chapter 2. Although water in the Imperial and Mexicali Valleys has achieved national agenda status during various periods of the twentieth century, the reactions of Washington and Mexico City to the topic have rarely been synchronized. The only exception is a 1944 bilateral treaty: In the unique context of the Second World War, Washington negotiated with state interests and signed with Mexico a formal agreement that would regulate the allocation of water from the Colorado River for both countries; Mexico obtained a minimum of 1.5 million acre-feet of water per year, and if a surplus of water existed during a particular year, Mexico could receive from the United States an allocation up to 1.7 million acre-feet of water

(Secretaría de Relaciones Exteriores 1975, 90–114; Worster 1985, 273, 321).

Because of the discontinuity in time in treating water as a security issue, this section will discuss the Mexican and U.S. cases separately. For Mexico, the creation of a buffer to the U.S. expansion from the mid-nineteenth century to the late 1930s and the salinity problem in the Mexicali valley in the 1960s were constructed as security issues. For the United States, in contrast, only the 1905–1907 floods became sufficiently salient to be treated as a security problem.

The Mexican Side

Borders are always worrisome territory for central governments, since they are remote from control and vulnerable to incursions from neighboring countries. This has certainly been the case with Mexico, which lost more than the 50% of its territory to the United States in the Treaty of Guadalupe-Hidalgo (1848) and the Gadsden Purchase (1853).

After 1848, Mexico's federal decision makers were concerned about the isolation of the Northern frontier and tried to avoid the "Texas syndrome" that would mean the loss of more territory to the United States. The fostering of new flows of population was precisely their formula to "integrate" such territory into Mexico. The Leyes de Reforma (Laws of the Reform) (1857) and the Decreto de Ley de Terrenos Baldíos (1863) (Decree on Untitled Land) were enacted with the purpose of bringing new settlers to isolated areas, including Baja California. According to their provisions, the government could sell untitled land to individuals or development enterprises at prices as low as 12 cents per hectare (Estrella 1982, 3). Unfortunately, abuses and frauds related to land ownership were common. In the subsequent years, land changed hands frequently. By the turn of the century, about 80% of the territory of Baja California was the property of American owners (Aguirre Bernal 1966, 99). The U.S. companies that acquired land and water in the Mexicali Valley favored the creation of a porous border that was clearly a threat to Mexican sovereignty (Grijalva Larrañaga 1988).

New security concerns arose in the Mexican federal government regarding the U.S.-Mexico frontier when Mexicali became a strategic site to many factions of the Mexican Revolution of 1910. In the context of the Mexican Revolution, the leaders of Magonismo, Enrique and Ricardo Flores Magón, sought to create a stronghold in this strategically valuable area, because it was isolated from the central government and close to the United States (Aguirre Bernal 1966, 91; Estrada Barrera 1978, 65). The Mexican federal government responded by sending troops to protect "the irrigation works of the Colorado River," and, paradoxically, the interests of the U.S. companies (Bonifaz de Hernández 1995, 369). Ultimately, the government defeated the rebels.

In the aftermath of the Revolution, Mexican policymakers decided to follow a strategy of negotiation rather than confrontation in dealing with the Colorado River Land Company, the most important foreign landowner in Mexico at the time. At first, postrevolutionary regimes in Mexico focused on winning formal recognition of their political legitimacy from the United States. When such recognition was obtained in 1920, these regimes were loath to alienate the Colorado River Land Company for a different reason. The governments of Álvaro Obregón and Plutarco Elías Calles feared that the expulsion of the company from Mexico would mean the end of the access to water from the Colorado River, which was the lifeblood of the Mexicali Valley. The search for security eventually brought an unanticipated outcome: The borderline became permeable for the expansion of private interests.

In the 1960s, Mexico again constructed the water in the Mexicali Valley as a security issue. At that time, the Mexicali area, with a population 300,000, had become a magnet for new immigrants from various areas of Mexico. Agriculture flourished, and cotton was the most important crop. There were, however, problems in its further development (Aguirre Bernal 1966, 437; Bernal 1996, 22). Water quality, rather than quantity, became the most serious threat to security in 1961. The Wellton-Mohawk irrigation district in Arizona dumped its agricultural wastewaters into the Colorado River for the purpose of draining this region of all the accumulated salinity. As a result, water delivered by the United States to Mexico contained elevated levels of salt, which ruined agricultural land in the Mexicali Valley (Bernal 1996, 3).

At first, the U.S. government did not consider salinity as a binational problem. Washington argued that the 1944 treaty related only to water quantity, not quality, and that Mexico was receiving the amount of water promised. Further, the United States believed damages were due to inadequate technical know-how in dealing with saline waters and deficient Mexican investment in drainage (Secretaría de Relaciones Exteriores 1975, 17–19).

In Mexico, grassroots movements fostered the interest of the federal government. In 1962, President Adolfo López Mateos declared that "the salinity issue is the greatest diplomatic problem confronting the two countries." In the early 1970s, President Luis Echeverría continued to press Mexico's claims through all available diplomatic channels (Fradkin 1981, 303, 308).

A bilateral agency, the International Boundaries and Water Commission (IBWC), as well as groups of ad hoc experts searched for solutions to the salinity problem. At the request of the federal governments in both countries, the Commission proposed certain actions that merely palliated the crisis without actually solving it. Minute 218 (IBWC 1965) stipulated that a canal be constructed to manage the waters from the Wellton-Mohawk district separately from good-quality waters from the Colorado River. Mexico thus would have the option of diluting high-salinity waters or diverting saline water to the ocean. Either way, Mexico was faced with greater insecurity in water supply, since nearly 75% of the waters from the Wellton-Mohawk were considered part of Mexico's allotment of water under the 1944 treaty (Secretaría de Relaciones Exteriores 1975, 154).

Under Minute 241 (IBWC 1972), the United States agreed to designate a special representative to find a solution to the salinity problem. By that time, Mexico had decided not to use the waters from the Wellton-Mohawk district and disposed about 122 million cubic meters of high-salinity waters directly into the ocean. The Nixon administration, particularly Secretary of State Henry Kissinger and former Attorney General Herbert Brownell, conducted a two-track negotiation with Mexico and with the Colorado-basin states (California, Arizona, Nevada, Utah, and Colorado) to move quickly toward an agreement on the salinity issue. The domestic negotiations were particularly difficult. The Colorado-basin states did not want to subscribe to an agreement "if only Mexicans were to benefit" (Fradkin 1981, 306, 316). They wanted federally funded salinity control projects in the upper basin as well as the construction of a desalination plant for water going to Mexico.

The final outcome of the bilateral U.S.-Mexico negotiations was the signing of Minute 242 by the IBWC in 1973, which stipulated the construction of a desalting plant, financed by the U.S. government, to treat the water from the Wellton-Mohawk district. In addition, this accord set a quality standard for water that Mexico would receive from the United States. Imperial Dam became the measuring point; the agreement required that waters above and below the dam have the same salt content. The security issue posed by salinity to Mexico encouraged a hard and definite definition of the border between the two countries.

As a consequence of these agreements, the salinity of water in the Mexicali Valley was reduced to 1,000 parts per million. Agriculture, however, never again became as successful economically as it once was. Rural areas never fully recovered. Even though Mexican government programs financed the construction of water canals and improved irrigation systems, the production of cotton diminished dramatically, with 1974 yields less than half those of 1959 (Bernal 1996, 25–26; Flores Caballero 1982, 126–128).

Even now, nearly thirty years later, the salinity problem in the Mexicali Valley is not totally solved. According to Mexican sources, the annual average index of the salinity of water sent by the United States complies with the norm established in 1973. Even so, monthly monitoring shows that during certain months of the year, salinity levels exceed the standard. However, local groups have been unable to raise salinity as a security issue and push it to a top diplomatic level again.

The U.S. Side

In the early twentieth century, water in the Imperial and Mexicali Valleys reached the status of a security issue in the United States. A technical mistake in the construction of a canal headgate in the Imperial Valley in 1905 resulted in the entire contents of the Colorado River being emptied into the adjacent valleys for more than two years. The resulting floods jeopardized agribusiness on both sides of the border. The Imperial Valley's crops were ruined, and Mexicali was practically destroyed. All the efforts to tame the river failed (Gottlieb and FitzSimmons 1991, 75; Worster 1985, 196–197).

The Theodore Roosevelt administration constructed the floods as a national disaster that posed a threat to security. Roosevelt allocated more than $3 million to the Southern Pacific Railroad Company to win "the battle against the Colorado." About 1,000 men with heavy equipment were dispatched to dump thousands of tons of gravel to restore the river embankment in the area where flooding had occurred. By early 1907, the first signs of success materialized (Callahan 1967, 462–463; Henderson 1968, 25). However, the danger from flooding due to events occurring in Mexico continued to plague U.S. policymakers. To diminish the risk of future floods, President William Howard Taft allocated funds to protect the lands and property in the Imperial Valley and other places along the Colorado River. Interestingly, the Taft administration implicitly acknowledged the binational nature of the irrigation works when it stipulated that the monies appropriated could be spent even on the Alamo Canal in Mexico, if necessary (Cory 1915, 1437; Estrella 1982, 9–10).

The serious consequences of the 1905–1907 floods marked the end of the consensus on the binational water arrangements that had existed up to that point. Previously there had been no serious criticism that the canal irrigating the Imperial Valley was built through Mexico. After the floods, strong voices in the United States sought to construct this arrangement as threatening national security and to redefine the border in a stricter and less permeable manner than previously. The proposition of constructing a canal through the U.S. territory—a project that U.S. governmental agencies and private companies had previously discarded because of its high cost—was resurrected (Fradkin 1981, 271). Discussion on the construction of such a canal continued up to 1928, when Congress finally approved the construction of the All American Canal. Water started flowing through this canal in full force in 1942.

Security constructions of water, like other constructions, have directly affected flows of people, their political mobilization, and their conceptions of the border between the United States and Mexico. As gatekeepers between the domestic and the international arenas, the governments of the two nations have left their mark when they perceived that national survival was in danger. The Mexican government promoted settlement of the Mexicali Valley in the name of security. Moreover, in the quest for development and population growth, Mexico's government relinquished sovereignty and control to U.S. private interests to the extent that the border between the two countries virtually disappeared. Later, the rising levels of salinity of waters delivered to Mexico limited agricultural employment and opportunity in the Mexicali Valley, with a direct impact on the settlement patterns of peoples. The salinity settlement of 1973 between the two nations imposed a definitive line demarking the border, since it defined the exact point at which salinity is measured. Constructing water as a security issue had an even greater impact on the U.S. side where security was the impetus for involving federal government agencies and for the construction of the All American Canal. Once the two countries no longer shared the canal system, the border between them became more fixed and less permeable.

Water as a Tool for Redistribution and Social Equity

At different times during the twentieth century, in both the United States and Mexico, water was envisioned as a tool to create more egalitarian capitalist agricultural societies in the Imperial and the Mexicali Valleys. Efforts to redistribute wealth from more to less well-off citizens was in each case directed from central governments operating in political climates favorable to agrarian reform. Local forces, including the new flows of people enticed to the valleys through newly created opportunities, however, were most influential in shaping the final outcomes. At the turn of the century, local forces in the United States solidly backed land reform, but once the U.S. entrepreneurial irrigation companies were vanquished, local interests preferred to run their own affairs. For the most part, local control and the sanctity of private property steered events away from the agrarian democratic ideal in both countries. In Mexico, after the social-oriented regime of Lázaro Cárdenas, Manuel Avila-Camacho's administration became the best ally of the local interests that preferred the creation of private properties over the *ejidos*—the core of the Mexican agrarian reform, a product of the Mexican Revolution of 1910.

The U.S. Side

The first U.S. settlers arrived in the Imperial Valley in January 1901. Private interests colonized the valley, and the first communities in the valley were not areas of equal opportunity (Taylor 1973, 4). Nevertheless, only the 1905–1907 floods interrupted the population growth in the valley. By 1909, the population numbered 15,000, and 160,000 acres were being irrigated, making the Imperial Valley by far the largest development in the Colorado basin (Fradkin 1981, 141, 186).

The Imperial Valley was colonized during an era when the U.S. government was actively attempting to shape the characteristics of the agrarian societies of the American West. The National Reclamation Act in 1902 was promoted by a coalition of interests that supported government intervention to support irrigated agriculture in the United States. The act's main purpose was to avoid monopolies on land ownership as well as the proliferation of absentee owners (Fradkin 1981, 275). The 160-acre limitation on land ownership per individual stipulated by the act was for years an issue of fierce national debate.

The U.S. Reclamation Service, created by the Secretary of the Interior after the National Reclamation Act was passed, was pivotal in altering the spirit of the 1902 legislation and bending it to very different ends from those for which it was originally intended, through the process of implementation. While carving its niche, this agency made alliances with the landholders in the West. In the Imperial Valley, the agency fought together with landholders against the dominance of the California Development Company (Gottlieb 1988, 44). The Reclamation Service supported the complaints of landholders against the company's inadequate irrigation works (which caused frequent floods) and its high water fees.

Alliance with local interests against the California Development Company came to supplant the original fervor of the Reclamation Service for redistributive land reform. By the late 1920s, the average size of land holdings in the Imperial Valley for the two principal crops (melons and lettuce) was 667 and 336 acres, respectively (Gottlieb and FitzSimmons 1991, 3; Fradkin 1981, 187)—well in excess of the Reclamation Act's 160-acre limit. Decade after decade, the Reclamation Service did little to change this situation. Amendments to this law to increase allowable acreage were continuously debated in the Congress. In 1928 the valley's landholders, the Imperial Irrigation District, and local politicians started a campaign to exclude the Imperial Valley from provisions regarding the 160-acre limitation as well as the residency requirements stipulated in the Reclamation Act (Fradkin 1981, 274, 276; Gottlieb and FitzSimmons 1991, 73). Then, the Secretary of the Interior Ray Lyman Wilbur "issued an exception" to the 160-acre limitation for the landowners in the Imperial Valley (Gottlieb 1996). In June 1980, the fifty-year debate on the 160-acre provision ended when the U.S. Supreme Court unanimously ruled in favor of the landholders' position (Fradkin 1981, 141, 287).

California's Imperial County is currently far from a model egalitarian agrarian community. Commercial agriculture with large holdings of land has replaced family farming. The original vision of creating a community of landholders residing in the valley has completely vanished (Gottlieb 1988, 97; Gottlieb and FitzSimmons 1991, 71; Fradkin 1981, 287; U.S. Department of Commerce 1994, 166). As a consequence, the towns located in the Imperial Valley remain small and relatively undeveloped, with machinery and landless migrant laborers (mainly of Mexican origin) performing most of the tasks once believed to be fundamental to agrarian democracy.

The Mexican Side
The proximity of Mexicali to the U.S.-Mexico border had served, from the aftermath of the Revolution of 1910 to the late 1920s, as a safety valve to take some of the steam out of demands for land by Mexican peasants. Mexican farm workers, who first settled in Mexicali, initially had the opportunity to cross the border and find a job in agriculture in the Imperial Valley or elsewhere in the United States. Things changed abruptly, however, during the Great Depression of the 1930s. Jobs on the other side of the border became scarce; thousands of Mexicans already working in the United States were deported. Many Mexican deportees settled in Mexicali and joined grassroots movements in favor of agrarian reform. They organized the *asalto a las tierras* (land takeover) and immediately contacted Mexican authorities to assert their right to land.

While pressures to end the control of the Colorado River Land Company escalated in the Mexicali Valley, President Lázaro Cárdenas launched an ambitious national agrarian reform, making the Mexicali Valley a major target. The reform included the creation of collective properties: the *ejidos*. (Aguirre Bernal 1966, 225–237; Sánchez-Ramírez 1990, 101–108).

In 1936, President Cárdenas came to an agreement with the Colorado River Land Company through which the company would sell its land to the Mexican government, which redistributed a portion of the land to Mexican colonizers. A year later, Cárdenas announced the redistribution of land purchased through the agreement to "all rural populations" settled in the Mexicali Valley before 1936 and granted land to about 4,000 prospective Mexican *colonos* (colonizers) (Sánchez-Ramírez 1990, 105–107).

Based upon the involvement of the state in the management of waters regulated by the Ley de Irrigación de Aguas Federales of 1926 (Federal Water Irrigation Law), the Mexican government established the *Distrito de Riego* (water district) of the Colorado River in 1938 (Walther 1996, 101–107). In addition, the construction of irrigation works became a priority in the federal budget's allocation to the Mexicali Valley (Bernal 1996, 23).

President Cárdenas perceived correctly that the U.S. government would not intervene in Mexico's agrarian reform. Theodore Roosevelt's administration made the cultivation of links with Latin America a priority of foreign policy. In addition, the Second World War caused the U.S. government to emphasize its "good neighbor policy" over the defense of certain U.S. private interests. The appeals of the U.S. owners of the Colorado River Land Company, aimed at convincing U.S. government leaders to persuade Mexico not to take over their lands, fell on deaf ears in Washington (Kerig 1988, 274).

Not many years after the *ejidos* were established, however, several *comisarios ejidales* (collective farm leaders) asked the Mexican federal government to create small and medium-sized private property in the Mexicali Valley. The redistribution of land to collective farms no longer had grassroots popular support. In 1942, the more conservative administration of President Manuel Avila-Camacho came into power and favored private property over *ejidos*. The new president gave individual 20-hectare plots of land to hundreds of farm workers. Since then, private properties have existed along with *ejidos* in the Mexicali Valley.

Mexican agribusiness flourished in the years following Avila-Camacho's privatization of land ownership in the Mexicali Valley. However, development was considerably limited from the 1960s on due to soil salinization in the valley. In addition, water quantity also became a problem. As the United States began to use all of its allocation of Colorado River water under the 1944 treaty, there was no excess to be allocated to Mexico per the 1944 treaty, and the flow of water to the Mexicali Valley was not sufficient to support the cultivation of certain crops. Further, as will be discussed later, water demand from urban areas was increasing.

The social purposes of reclamation and agrarian land reform are all but forgotten today in both the United States and Mexico. In the United States, the reclamation movement consolidated the Imperial Irrigation District's control over U.S. water and eroded the potential for binational or integrated water management. The U.S. side perceived the pattern of land reform chosen by the Cárdenas government as so "socialistic" and foreign as to further dilute any previously existing feelings of joint dependence on a shared Colorado River.

The ultimate failure of democratic agrarian reform established the preconditions for a redefinition of water as a commodity. Had thickly populated communities of family farmers with roughly equal status existed in the rural areas of Mexico, the challenge of market transfers of water to urban areas from rural areas would have met with greater resistance.

Water as a Commodity

Much currency is accorded to the idea that water can be managed in a more productive and efficient manner when viewed as "a tradable standardized commodity" rather than as a product of engineering or an in situ part of nature. This view of water prevails in both Mexico and the United States today; it has permeated governmental agencies and has generated support among a number of different segments of the population in both countries. As suggested in chapter 2, markets serve as the conduits for moving water from lower- to higher-value uses, perhaps in different geographical jurisdictions, such as across international borders or in cities rather than in the countryside. Moreover, the notion of allowing water to flow in the direction of whoever is willing to bid the most for it, even if that means allowing the nation's water to escape its control, is hardly unique to the Imperial and Mexicali Valleys.

Turning first to the Mexican experience, population growth in Mexico became an important factor in water management by the early 1960s. Water acted as a magnet drawing people to the Mexicali Valley, first in agriculture and later in other pursuits. New waves of immigrants had come to the valley from other regions of Mexico. Many immigrants hoped that the supply of water that had generated commercial agriculture would also provide them jobs. In addition, many Mexicans came to Mexicali from points south to participate in the Bracero Program, a U.S.-Mexico labor agreement implemented from 1942 to 1964 that facilitated agricultural work for Mexican citizens in the United States.

It is not surprising that the transfer of water from rural to urban settings began earlier in Mexico than in the United States, since Mexican development policy explicitly acknowledged that greater economic returns were likely to come from industrialization and moving people from rural areas to become urban laborers. During the 1960s, the population of Mexicali grew steadily in spite of the problems of salinity and declining agriculture that the region faced (table 3.1). In the midst of growing labor supply and shrinking labor demand in agriculture, employment prospects in Mexico were brightened by the Border Industrialization Program (BIP). Initiated in 1965, the BIP allowed foreign companies to own and operate manufacturing plants (*maquiladoras*) in Mexico using Mexican labor. The first *maquiladora* in Mexicali opened its doors in 1966; by 1995, 160 *maquiladoras* were in operation there (Ingram et al. 1995; Jenner 1992, 48–49).

Currently, 92% of the people in the Mexicali Valley reside in urban areas. In the 1990s, Mexicali City consumed 15% of the total water demand of the valley, up from 8% in the 1980s (Cortez 1997, 3). Meanwhile, agriculture in Mexicali used 85% of the water, although only 8% of the population resided in the rural areas (Bernal Rodríguez 1991, 229;

Table 3.1
Population growth in the border: Baja California and California, 1930–1995

Year	San Diego	Imperial Valley	Mexicali Valley
1930	209,659	60,903	29,985
1960	1,033,011	72,105	281,333
1990	2,498,016	109,303	601,938
1995	2,690,255	141,500	695,805

Sources: California Statistical Abstract (1995), Census of Population, State of California (1990), and XI Censo General de Población y Vivienda (1995).

Cortez 1997, 3). As water demand by the urban sector increases, so does sectorial competition for water from the Colorado River in the Mexicali Valley. The population in the Imperial Valley has also been increasing since the 1960s, albeit at a slower rate than that of Mexicali (see table 3.1). In 1995, 141,500 people lived in the Imperial Valley. Agriculture in Imperial Valley is characterized by mechanization and a small percentage of the population works in agriculture (about 1,200 people or less than 1%) (State of California 1997; Gerber 1997, 1; Whiteford and Cortez 1996, 133). In contrast to the position of agriculture in Mexico, however, the Imperial Valley's 500,000 acres of farm land are still considered a key agricultural area in the United States.

In both the Mexicali and the Imperial Valleys, contemporary water transfers from rural to urban areas have been designed to take place within (but not between) each nation. However, viewing water as a commodity has caused the border to become once again, from time to time, somewhat soft or permeable. For example, one incentive Mexico has used to encourage foreigners to establish *maquiladoras* is subsidization of the price of water. Although the in-bond plants (or *maquiladoras*) compete with agriculture and urban dwellers for the good-quality water available in the Mexicali Valley (Jenner 1992, 49), the Mexican government has not hesitated to use its considerable authority over water allocation to favor urban over rural water users.

The favoritism toward urban water uses reflected in the actions of the Mexican government is reinforced by changes in law that encourage the treatment of water as commodity. From the aftermath of the Mexican Revolution until 1992, a federal agency (currently the Comisión Nacional del Agua or National Water Commission) and federally established irrigation districts had absolute power over water allocation. Barter or sale of individual water rights was not legal in Mexico. The reform of Article 27 of the Mexican Constitution, and the passage of the Ley de Aguas Nacionales (National Waters Law) in 1992 dramatically changed the long-standing tradition of the Mexican state as the ultimate owner of water in Mexico and modified the long-standing link between land and water. According to the 1992 law, water now can be sold or purchased independently of the possession of land, and markets for water can be created. To date, water has only been transferred from landowner to landowner within a single *módulo* (subdivision of an irrigation district). However, due to growing demand in the urban sector, it is likely that the *módulos*, particularly those closer to the city of Mexicali, will soon begin selling water to urban areas (Cortez 1997).

Since the 1980s and through the 1990s, an environment supportive of the creation of water markets has also prevailed in the United States. As all of the existing surface and groundwater supplies by this time had become fully appropriated and even overutilized in southern California, thirsty cities looked toward agricultural supplies as an alternative source. A web of local, state, and national agencies have been involved in the transfer of water from the Imperial Valley to urban areas of Southern California.

In 1989, the Metropolitan Water District (MWD), which spans the Los Angeles/San Diego urban corridor, paid the Imperial Irrigation District (IID) $200 million to fund water conservation projects in the Imperial Valley, in exchange for 106,000 acre-feet of water turned over annually to MWD for the next thirty-five years (McClurg 1996, 7). Currently, the IID and the San Diego County Water Authority (SDCWA) are involved in what promises to be the largest rural-to-urban water sale in U.S. history. If the deal is successfully concluded, it will most certainly usher in a new era of water marketing and commodification of water throughout the Colorado River basin. Lured by the possibility of gaining substantial revenue from the deal, the IID decided to sell part of its 3.3 million acre-feet annual allocation of Colorado River water rights to the SDCWA. However, the cost of the water to be transferred is still a matter of dispute between the SDCWA and the MWD. If the agreement that was signed in April 1998 (SDCWA 1998) is fully implemented, the water will be transferred to MWD for forty-five years. In the first year of the agreement, 20,000 acre-feet will be delivered. The amount of water will increase every year up to a maximum of 200,000 acre-feet per year (SDCWA 1997b).

As in Mexico, the U.S. federal government, specifically the Department of the Interior (DOI), to which the Bureau of Reclamation (the successor to the Reclamation Service) reports, and which is in charge of water matters in the West, has viewed the IID-SDCWA deal with enthusiasm. The DOI has described this agreement as a commendable model for intrastate transfers of the sort that could solve the problem of increasing demand for water in urban California. The state government may play a more active role in facilitating the transfer by pledging $200 million from a prospective $1 billion water bond issue to facilitate the agreement between the contending parties (Rohrlich 1998, 7).

The IID-SDCWA water transfer, should it come to fruition, may establish a model even for international water transfers. For example, to transport the water that the SDCWA intends to purchase from the IID, the SDCWA, in cooperation with Mexico, is studying the building of a parallel facility to the Colorado River Aqueduct, which transports water from the Colorado River into California (CVWD 1996; SDCWA 1997b). Although transnational water sales to fuel *maquiladora* development have not yet taken place, in other ways, the fostering of urban growth through the commodification of water has operated to make borders more porous and permeable. As discussed earlier, California business interests once controlled the company serving water to the Mexicali and Imperial Valleys during the heyday of irrigated agriculture, and California interests now control many *maquiladoras*. In some real sense, Mexicans have lost control over their own lands and futures or, at least, they share that control with U.S. businesses.

The conceptualization of water as a commodity has been portrayed as positive in both the Mexicali and Imperial Valleys. Among the benefits cited is an increase in water conservation programs. Supporters also argue that the transfer of water derived from conservation efforts can pump millions of dollars into water-selling rural economies (SDCWA 1997).

Despite these arguments, many rural residents have resisted water transfers but generally have not had much opportunity to voice their concerns. In Mexico, the design of water markets has been a top-down (or government-designed) activity. Cortez (1997) reports that in 1992, 57% of the rural dwellers in the Mexicali Valley were in fact opposed to any modification of the Mexican Constitution. Most were afraid that their *módulo* could not afford both to maintain the infrastructure and to conduct further public works if the Constitution was modified, and others had doubts about the democratic procedures within the *asociaciones de usuarios* (users' associations). However in 1998, a mere six years later, surveys indicated that opposition to the new arrangements had decreased to 31% (Cortez-Lara and García-Acevedo 2000). The users now acknowledge the advantages of getting rid of the red tape that characterized the Comisión Nacional del Agua. However, the situation within each of the *módulos* varies in terms of land holdings and access to water (Bernal Rodríguez 1991, 29).

On the U.S. side, land in the Imperial Valley has been concentrated in the hands of large absentee landowners who are attracted to the Imperial Valley for its income-generating potential. These landowners have been enthusiastic brokers for the water deal with the SDCWA (SDCWA 1997b). The rising price of water provides an opportunity for profit making but does not attract residents with commitments to the long-term welfare of the Imperial Valley as a human community (Gottlieb 1998, 269).

Among those who perceive themselves as potential losers from water transfers in the United States are some of the residents in the small towns of the Imperial Valley, as well as some small and medium-sized landholders who feel their already difficult positions will get worse. The community may become polarized into factions: those who will profit from the sale of water and those who will not. Because less water will be available to the Imperial Valley for agriculture and other economic uses, unemployment may increase in the Imperial Valley, already within the poorest county of the state of California (in terms of personal income).

It is not at all clear that rural residents in the Imperial and Mexicali Valleys entirely subscribe to the notion that water is merely a commodity. For many small townspeople and rural residents, water has a more symbolic meaning as a way of life in which community, water, and land are inextricably tied together. Other, increasingly vocal groups are contesting the commodity view of water, but from a more environmental perspective. A network of transnational organizations, importantly including the International Sonoran Desert Alliance, constructs water in ecological terms, viewing the Colorado River as part of the upper gulf of the Sea of Cortez, among the richest biological treasures in the world (Yetman 1996; Brown 1994; West 1993). For some of these activists, the ultimate aim is the creation of an international biosphere reserve that would protect both biota and indigenous people from the harmful side effects of border economic development. Although in its very early stages, the grassroots movement spans borders and cultures and is not fundamentally dissimilar to the Yukon to Yellowstone initiative described by Suzanne Levesque in chapter 5.

Conclusion

This chapter has demonstrated the insights into the changing meanings of water that can be gained through an in-depth historical case study. Without the perspective this longer view provides, the optimistic proclamation of the Secretary of the Interior that "the focus of river management has now shifted toward managing our supply for greatest efficiency and productive use" (McClurg 1996, 4–5) might well be received uncritically. The historical review provided in this chapter of conflicting and overlapping meanings of water as they have been adopted in different eras suggests that the commodity view of water may well be transitory. Water may become the focal point for cross-border alliances that challenge the value premises of the commodity view. A fundamental value conflict between developmental and environmental advocacy coalitions could ensue.

A century of experience in manipulating water as property, product, and commodity suggests that water decisions are much less socially benign than modern portrayals suggest. The viewing of water as a commodity will certainly increase the disparities between the United States and Mexico and the polarization between rural and urban settings within each country and may exacerbate the cleavages between the rich and the poor in both countries. Moreover, struggles may well develop between one urban area and another for water to support continued growth. These conditions set the stage for a redefinition of water to replace the prevailing commodity view.

The close association of water, population, and the character of the border will continue, however, regardless of how water may next be construed. Explicit recognition of the linkage of water, people, and the border in public discourse might well contribute to avoiding the unanticipated adverse impacts that various constructions of water have imposed.

References

Aguirre Bernal, Celso. 1966. *Compendio histórico-biográfico de Mexicali.* Mexicali, Mexico: Talleres de Litografía Educativa.

Aguirre Bernal, Celso. 1995. "La colonización de las aguas del Río Colorado." Unpublished manuscript.

Alvarez de Williams, Anita. 1974. *The Cocopah People.* Phoenix, Ariz.: Indian Tribal Series.

Alvarez de Williams, Anita. 1975. *Travelers among the Cucapa.* Los Angeles: Dawson's Book Shop.

Alvarez de Williams, Anita. 1997. "People and the River." *Journal of the Southwest* 39, nos. 3 and 4 (Autumn–Winter): 331–351.

Bernal, Francisco A. 1996. "Tenencia de la tierra y recursos hidráulicos en la región fronteriza México-Estados Unidos: El Valle de Mexicali." Paper presented at the seminar "Who Owns America? Land and Resource Tenure Issues in a Changing Environment," University of Wisconsin–Madison.

Bernal Rodríguez, Francisco. 1991. "Análisis de las eficiencias en el uso y manejo del agua de riego en el Valle de Mexicali." In José Luis Trava Manzanilla et al. (eds), *Manejo ambientalmente adecuado del agua: La frontera México–Estados Unidos.* Tijuana, Mexico: El Colegio de la Frontera Norte.

Bonifaz de Hernández, Roselia. 1995. "Los Sucesos de 1911." Unpublished paper. Available through the library of El Colegio de la Frontera Norte, Tijuana, Mexico.

Brown, David E., ed. 1994. *Biotic Communities, Southwestern United States and Northern Mexico.* Salt Lake City: University of Utah Press.

Callahan, James Morton. 1967. *American Foreign Policy in Mexican Relations.* New York: Cooper Publishers.

Castetter, Edward F., and Willis H. Bell. 1951. *Yuman Indian Agriculture.* Albuquerque: University of New Mexico Press.

Checcio, Elizabeth, and Bonnie G. Colby. 1993. *Indian Water Rights: Negotiating the Future*. Tucson: University of Arizona Press/College of Agriculture.

Cocopa Indian Tribe Website ⟨www.azcentral.com/depts.aaliving/indians⟩.

Cortez, Alfonso. 1997. "Solución al conflicto del agua. A campo abierto." *Mexicali* 2 (January–February): 3–5.

Cortez-Lara, Alfonso, and Maria Rosa García-Acevedo. 2000. "The Lining of the All American Canal: The Forgotten Voices." *Natural Resources Journal* 40, no. 2 (Spring): 261–279.

Cory, H. T. 1915. *Imperial Valley and Salton Sink*. San Francisco: John J. Newbegin.

Coachella Valley Water District (CVWD). 1996. "California Basic Colorado River Water Entitlement 4.4 Million Acre Feet" ⟨www.cvwa.org/wateriss/colorado/html⟩, July 1.

Estrada Barrera, Enrique. 1978. *El Río, cronología de Mexicali*. Mexicali, Mexico: Author.

Estrella, Gabriel. 1982. *El Orígen de la región de los valles de Mexicali e Imperial*. Mexicali, Mexico: Universidad Autónoma de Baja California.

Flores Caballero, Romeo. 1982. *Evolución de la Frontera Norte*. Monterrey, Mexico: Centro de Investigaciones Económicas, Universidad Autónoma de Nuevo León.

Fradkin, Philip L. 1981. *A River No More*. New York: Alfred A. Knopf.

Gerber, Larry. 1997. "California Cities' Farmers Clash over Imperial Valley Produce Insatiable Thirst." *Los Angeles Times*, June 1, B-1.

Gottlieb, Robert. 1988. *A Life of Its Own: The Politics and Power of Water*. San Diego, Calif.: Harcourt Brace Jovanovich.

Gottlieb, Robert, and Margaret FitzSimmons. 1991. *Thirst for Growth: Water Agencies as Hidden Government in California*. Tucson: University of Arizona Press.

Grijalva Larrañaga, Edna-Aidé. 1988. "Primeras entradas al Valle de Mexicali." In Michael Mathes (ed.), *Baja California*. México: Instituto de Investigaciones Mora.

Henderson, Tracey. 1968. *Imperial Valley*. San Diego, Calif.: Neyenesch.

Ingram, Helen, et al. 1995. *Divided Waters: Bridging the U.S.-Mexico Border*. Tucson: University of Arizona Press.

International Boundary and Water Commission, United States and Mexico. (IBWC). 1965. Minute 218, Recommendations on the Colorado River Salinity Problem. March 22. Ciudad Juarez, Chihuahua, Mexico.

International Boundary and Water Commission, United States and Mexico. (IBWC). 1972. Minute 241, Recommendations to Improve Immediately the Quality of Colorado River Water Going to Mexico. July 14. El Paso, Texas.

Jenner, Stephen R. 1992. "Free Trade and Border Industry." In Paul Ganster and Eugenio O. Valenciano, *The Mexican-U.S. Border Region and the Free Trade Agreement*. San Diego, Calif.: San Diego State University.

Kelly, William H. 1977. *Cocopa Ethnography*. Tucson: University of Arizona Press.

Kerig, Dorothy P. 1988. *Yankee Enclave: The Colorado River Land Company and the Mexican Agrarian Reform in Baja California*. Ph.D. diss., University of California, Irvine.

McClurg, Sue. 1996. "Colorado River Controversies." *Western Water* (March–April): 4–13.

Morrison, Jason, et al. 1996. *The Sustainable Use of Water in the Lower Colorado River Basin*. Oakland, Calif.: Pacific Institute for Studies in Development, Environment, and Security.

Rice, John. 1997. "Where the River Runs Dry, a Torrent of Change, Pain Follow." *Los Angeles Times*, June 1, B-3.

Rohrlich, Ted. 1998. "Thirst to Overhaul Powerful Water Agency Grows Stronger." *Los Angeles Times*, July 19, B-1.

Sánchez-Ramírez, Oscar. 1990. *Cronica Agricola del Valle de Mexicali*. Mexicali, Mexico: Universidad Autonóma de Baja California.

San Diego County Water Authority (SDCWA). 1997a. "San Diego Imperial Valley Water Transfer Would Benefit California, Chairwoman Says." SDCWA Website: ⟨www.sdcwa.org/text/pressrel/frahmiid.htm⟩. March 18.

(SDCWA). 1997b. "Fact Sheet: Water Conservation and Transfer Agreement San Diego County–Imperial Valley." SDCWA Website: ⟨www.sdcwa.org/text/dfaq.htm⟩. December 1.

(SDCWA). 1998. "Landmark Water Conservation and Transfer Agreement Ratified." SDCWA Website: ⟨www.sdcwa.org/text/pressrel/dvote2.htm⟩. April 29.

Secretaria de Relaciones Exteriores. 1975. *La Salinidad del Rio Colorado*. Mexico City: Coleccion del Archivo Diplomatico Mexicano.

State of California. 1997. *California Statistical Abstract*. Sacramento, Calif.: Author.

Taylor, Paul S. 1973. "Water, Land and Environment, Imperial Valley: Law Caught in the Winds of Politics." *Natural Resources Journal* 13, no. 1 (January): 1–35.

U.S. Bureau of the Census. 1998. *Statistical Abstract of the United States* (118th ed.). Washington, D.C.: Government Printing Office.

U.S. Department of Commerce. 1994. *1992 Census of Agriculture. California. State and County Data*. Washington, D.C.: Government Printing Office.

Walther, Adalberto. 1996. *El Valle de Mexicali*. Mexicali, Mexico: Universidad Autónoma de Baja California.

West, Robert C. 1993. *Sonora: Its Geographical Personality*. Austin: University of Texas Press.

Whiteford, Scott, and Alfonso Cortez. 1996. "Conflictos urbano rurales sobre el agua del Río Colorado en el ambito internacional." In *Agua: Desafíos y oportunidades para el Siglo XXI*. Aguascalientes, Mexico: Gobierno del Estado de Aguascalientes.

Worster, Donald. 1985. *Rivers of Empire: Water, Aridity and the Growth of the American West*. New York: Pantheon Books.

Yetman, David. 1996. *Sonora: An Intimate Geography*. Albuquerque: University of New Mexico Press.

4

Lessons from Lake Constance: Ideas, Institutions, and Advocacy Coalitions

Joachim Blatter

The cooperative effort to protect Lake Constance (Bodensee) is recognized worldwide as one of the first and most successful examples of international environmental regimes (Breitmeier et al. 1993; Scherer and Mueller 1994).[1] Whereas the lake was close to a biological collapse in the 1960s, thirty years later its waters are being sold as high-quality table water. The fact that such an environmental success story could have been achieved with respect to a watercourse even though the riparian states of Germany, Switzerland, and Austria never formally agreed on a clear-cut boundary on the lake is not the only puzzling aspect of this transboundary cooperation. The following case study of the regulation of boats on Lake Constance shows that a postmodern cross-border environmental politics deviates in many ways from the conventional modern concepts of international (environmental) regimes.

Although the cross-border cooperation outlined in this chapter does not typify cross-border water politics in most regions of the world, the case of Lake Constance does serve as an instructive look into a possible future for other regions. The problem of sport boats on Lake Constance is a by-product of growing (albeit unequally distributed) material affluence and the search for self-realization through recreational activities—characteristics that are becoming more typical throughout the world. Today, it is highly uncommon for cross-border politics to be conducted in an institutional environment marked by three federal nation-states and a thoroughly institutionalized political-administrative system at the continental level (the European Union). Such a situation will become, without doubt, more common in the future, as politics move toward "glocalization" (Robertson 1995), "intermestic politics"

(Manning 1977), and "multi-level governance" (Marks, Hooghe, and Blank 1996).

It is particularly important to this volume that the Lake Constance case illustrates the ways in which contemporary transboundary water governance deviates from a number of modern assumptions based on narrow notions of rationality and the predominance of nation-states as principal actors. The following propositions will be explored in sequence.

Transboundary cooperation at Lake Constance is not well explained by the existence of "problematic situations" in which actors attempt to avoid negative externalities and capture positive externalities (Zürn 1992). Instead, specific contextual factors, such as the framing of issues and institutional competition to capture political credit, instigate and channel action.

International systems are increasingly differentiated into many different arenas of interaction through which actors' preferences are expressed. The number of subnational transboundary networks and alliances has grown enormously in recent decades, as chapter 1 suggests. The "two-level game" (Putnam 1988) or "two-stage process" (Moravcsik 1997) in which nation-states are pivotal institutions for subnational interest aggregation prior to international negotiations no longer captures reality. Instead, transnational advocacy coalitions made up of subnational public officials and voluntary associations directly express sectoral interests, which other regional institutions then mediate.

Lake Constance: Problems and Regulation of a Cross-Border Recreational Region

A Short Characterization of Lake Constance

Covering an area of 539 square kilometers, Lake Constance is the second-largest lake in central Europe after Lake Geneva. Of the 265-kilometer shoreline, 52% belongs to the German Bundesland (subnational state) Baden-Württemberg. Barely 10% of the shoreline belongs to Bavaria, and 11.5% belongs to the Austrian Bundesland Vorarlberg. The Swiss shoreline is divided between the cantons of St. Gallen (4.2%)

and Thurgau (23%). Although not bordering the lake directly, the Swiss canton of Schaffhausen, located just beneath the mouth of the Rhine where it drains water from Lake Constance, is also considered a member of the region.

The only region in Europe in which the borders of the neighboring countries were never formally delineated, Lake Constance is "a curiosity in international law" (Orbig 1990, 376). Neither the Westphalian Peace Treaty of 1648 nor any subsequent accord contains any kind of legal requirements for settling border issues on the lake. The smaller littoral states and cantons (in particular) consistently defend the condominium theory, which (in contrast to the real separation theory) declares that the greatest part of the lake constitutes a common sovereign territory (Graf-Schelling 1978, 22–24).

Since environmentalist opposition thwarted a plan to make the upper Rhine navigable from Basle to Lake Constance during the 1960s, Lake Constance has to this day primarily served as a storehouse for potable water and the regional focal point of tourism. The beginning of the extensive use and far-reaching distribution of the water of Lake Constance reaches back more than 100 years, when in 1895 the Swiss city of St. Gallen put a waterworks into service in Rieth, near Goldach. Other cities followed this example and today, from eighteen removal stations, about four million users annually consume approximately 180 million cubic meters of water from the lake.

In 1908, the institution of "Lake Constance Week" marked the first peak of the lake's usage for water sports and tourism. Lake Constance Week was intended to raise Lake Constance to a pivotal location in the international motorboat racing community (Trapp 1994, 24–29). Although in the first half of the century (motored) water sports remained reserved for a small class of wealthy people, they became mass entertainment amidst the increasingly affluent, leisure-oriented society that arose following the World Wars.

The meanings of water held by the civil actors involved in Lake Constance water governance vary according to their interests. In terms of the meanings of water denoted in chapter 2, water is variously constructed as a cultural specific good (by boaters, water sports enthusiasts, and the tourism industry) and as a natural specific good (by environmentalists, ornithologists, and bird watchers). As we will see in the following sections protection of the water of Lake Constance also became the central symbol for politicians around the lake in their attempts for building a cross-border "Euroregion."

Pollution of the Lake by Boats and Ships

The ecological problems created by boating on Lake Constance are attributable to the sheer number of boats. From the end of the Second World War until the close of the 1970s, the number of boats on the lake rose sharply each year. In the 1980s the numbers continued to grow, but not as quickly. At the beginning of the 1990s, the total count of authorized boats on Lake Constance stabilized around 55,000. A trend toward the use of increasingly powerful motors is also evident (Internationale Gewässerschutzkommission 1994).

The extent to which boating pollutes Lake Constance is contested. The following section will explain the different pollution factors that stand in the center of the debate. Two basic forms of lake pollution by shipping can be distinguished: structural (ecosystemic) damage and material (chemical) pollution.

Structural damage to the lake by ships and boats includes the destruction of habitats as well as the disruption of other recreational activities. The enormous exhaustion of land by harbors, landing stages, and buoy fields weigh especially heavily. Of the 160 kilometers of the Baden-Württemberg lake shoreline, shipping facilities blanket a total of 45 kilometers. All of these facilities lie in the important shoal area and cause specific damage or the total destruction of shoals. The shoals closest to the shoreline act as contact zones between land and water; not surprisingly, they are especially diverse environments with large numbers of species. Moreover, these areas clean the lake through the gradual elimination of organic substances (Ministerium für Ernährung, Landwirtschaft, Umwelt und Forsten Baden-Württemberg 1981; IGKB 1987). Shipping facilities not only destroy lake shoals but also exhaust valuable natural environments on land.

In addition to the evident damage wrought by shore facilities, boats also impact nature by startling waterfowl and by producing waves.

Damage to the ecosystem of Lake Constance has far-reaching consequences for all of Europe. It is the migrational resting spot and winter quarters for 300,000 waterfowl from northern and eastern Europe as well as western Siberia. When disturbed during their rest, these birds cannot regenerate sufficiently for their long migrational flight (Deutsche Umwelthilfe 1991). The waves caused by fast boats are also blamed for massive decline of the lake's reed stands. Last, but not least, noise pollution from motorboats on the lake and their land transport upsets visitors seeking relaxation.

Poisonous to water organisms even in relatively small concentrations, hydrocarbons emitted by motorboats stand center stage in terms of chemical damage attributable to boats on the lake. According to the (controversial) calculations of the Internationale Gewässerschutzkommission (International Commission for the Protection of Lake Constance) (IGKB 1982, 16), boating emissions in 1980 amounted to 1,120 tons of light hydrocarbons, of which 755 tons came from sport boating, and 42 tons of heavy hydrocarbons (21 tons from sport boating). Other harmful substances, such as carbon monoxide, sulfur dioxide, lead, and nitric oxide, are introduced into both water and air (IGKB 1982, 18).

More problematic chemicals are introduced into Lake Constance by materials (phosphates) used for cleaning boats and through so-called antifouling material (special paint, containing chemical compounds that make cleaning boat bottoms easier). Two important aspects in regard to this damage should be highlighted. First, the structural damage blamed on shipping affects foremost the shorelines of the national regions along the lake; the chemical damage, on the other hand, affects the common water. Second, the amount of chemical damage is estimated on the basis of the number of the authorized boats. It is very much contested and, to this point, has not led to a real detrimental effect on the quality of the drinking water.[2] Conversely, structural damage is ascertained through direct measurements and observation.

Steps and Measures to Regulate Boats on Lake Constance

Political activities to regulate boats and ships on Lake Constance began in the 1960s after environmentalists thwarted plans to open the River Rhine to shipping from Basle to Lake Constance. Simultaneously, calls to limit boats arose in all of the neighboring states; among the most strident demands were bans on motorboats and new landing slips. Also during the 1960s, conferences were organized to revise the International Shipping and Harbor Rules of 1867. After an intensive political discussion, the Bodensee-Schiffahrts-Ordnung (BSO) (Lake Constance Shipping Regulations) was passed in 1973. Instead of the stringent boating restrictions repeatedly requested, only a ban on two-cycle motors over 10 horsepower was passed. Most motorboats using the lake at the time already had modern four-cycle motors or motors under 10 horsepower, so the ban did not have a large effect on damage to the lake from motored boating.

In the 1980s, after a report by water conservationists detailed problems caused by boats, a second round of cross-border efforts to regulate boats on the lake began. Once again, drastic measures against the boats were placed on the agenda. As these demands proved untenable, a search began for a technical solution. After long discussions and many technical investigations, exhaust specifications for boats on Lake Constance were finalized in 1991. The first stage of the exhaust regulations was put into force on January 1, 1993; the second stage, whose mandates surpass the general regulations of the European Union and Switzerland, took effect in 1997. This second effort, which imposed strict emission standards, spurred technological innovations in construction of boat motors. Nonetheless, stricter measures against structural damage, such as reducing the number of locations for boat landings, were strongly resisted by boat owners.

How can the development and success of cross-border cooperation in regulating boating on Lake Constance as well as the environmental political results now be explained? The following discussion demonstrates that transboundary action cannot be explained as the recognition of a problematic situation in which rational actors respond to self-interest. Nor does the modern notion of interest aggregation at the national level, and the expression of that nation-state interest at a second international level or process, fit the reality of water politics at Lake Constance. As chapter 1 suggests, an examination of linkages and alliances is necessary to develop an understanding of water-related conflicts and their resolution.

Inadequacy of Modern Approaches: Does the Problem of Boats on Lake Constance Represent a Compelling Reason for International Cooperation?

Is the problem of boats on Lake Constance consistent with a "collective-action" problem (Ostrom 1990), in which spillover effects of positive and negative externalities necessitate the construction of cross-border institutions? At first glance the boats do seem to pose a collective-action problem: To a great extent, the boats move on water that is not allocated to one country. A closer look, however, reveals that significant cross-border interdependence exists only in regard to one material problem: toxic pollution of the water. The water cannot be assigned clearly to one side; thus, pollution from a German boat threatens the quality of the drinking water for the Swiss. This alone is not sufficient to explain action, however, and does not correlate well with measures actually taken.

Structural damage to the lake from boats shows that the littoral states are confronted with an "equal problem" but not a "common problem."[3] Structural damage to the flora and fauna of the lake arises mostly through the landing docks and through the transport of people and boats to the lake. This damage is locally determined and therefore easily assigned to a specific jurisdiction. For example, Swiss boats damage German shoals only marginally. Thus, all sides face the problem of structural damage by boats, but the damage to each develops almost exclusively through its own activities. For shoal damage, then, a compelling necessity for cross-border coordination of regulation does not exist.

The emergence of a natural specific conception or symbolization of water in Lake Constance, rather than some territorially specific interest, provides a far better explanation of the basis of action taken to regulate boats on the lake. Only an ecosystemic perspective on Lake Constance suggests a certain interdependence of the regulation of boat landings. For instance, the biochemical cleaning work of the shoals benefits the entire lake. But the greatest ecological value of Lake Constance lies in its habitat function as a springboard for the far-reaching migration of birds. Cross-border cooperation on the basis of an ecological interdependence compels cooperation with the regions of northern Europe, where the birds rest in summer, but it is not an absolute necessity for regional cross-border cooperation.

Based on this analysis, logic suggests that postmodern nature-based and nonmaterial concerns far outweighed the material problem of chemical pollution in determining the measures taken to address water quality problems. Positing the threat of chemical pollution of the commons as a motivator for action taken to regulate boating is subject to too many challenges. First, why did the strongest demands for a common regulation come about when data on chemical pollution were still unavailable? In addition, the scientific evidence that motorboat emissions represent a real threat to water quality in the lake is far from compelling. Second, why does the political cross-border cooperation concentrate primarily on the material emissions of sport boats, as their impact is relatively minor compared with other sources? As the next section illustrates, a look at the general discourses and at the institutional setting at the time of the regulatory breakthroughs shows that competition among institutions to capture the framing of issues and to establish institutional legitimacy provides a more compelling explanation of what transpired.

The Discovery and Framing of the Problem by Cross-Border Institutions

If the structural damage by boats does not present a cross-border problem, and if the chemical pollution cannot be seen as an unquestionably relevant problem, how can it be explained that the theme of cross-border boat regulation has taken a very prominent position in the political agenda of the Lake Constance region for thirty years? Since the initiation of cross-border cooperation can be explained only partly by functional necessity, contextual factors are illuminated here to supplement this explanation, thus contributing to a more accurate understanding of the process of cross-border governing.

The emergence of ecological problems attributable to boat use on Lake Constance in the 1960s, at a time when the number of boats was substantially lower than today's number, can be explained by the concurrent institutionalization and professionalization of nascent environmental organizations. Widely mobilized because of their successful campaign against the proposal to make the Rhine navigable between Basle and Constance, environmentalists shifted their focus in the mid-1960s to other, similar problems. At this time, it was primarily ornithologists and birdwatchers among the environmentalists who pointed to the structural damage caused by shipping on the lake. From their perspective, the continuously increasing number of sport boats on Lake Constance posed a threat to valuable habitats that equaled the ecological threat of the proposed extension of shipping on the Rhine between Basle and Lake Constance.

In the early 1970s, discussion about regulation of boats on the lake first culminated in intense cross-border discussions and negotiations. A familiarity with the historical context is requisite to understanding the cross-border activities on Lake Constance (see table 4.1). At the beginning of the 1970s, the Council of Europe (Europarat) spurred a wave of initiatives for cross-border cooperation.[4] First, the council dubbed 1970 the "European Year of Nature Protection," inspiring considerable momentum toward regular cooperation among environmental groups. The Arbeitsgemeinschaft Naturschutz Bodensee (ANU) (Study Group for Environmental Protection of Lake Constance), a loose umbrella organization founded at that time, unified thirty-three private nature protection groups and citizen initiatives encompassing 18,000 members in the Lake Constance region (interview with Harold Jacobi, Deutsche Umwelthilfe, June 28, 1994).

Even more significant was a second initiative of the council that made cross-border cooperation a relevant topic for politicians and governments. The first "European Conference of the Ministers of Spatial Planning" in September 1970, initiated by the council, passed a recommendation to set up cross-border spatial planning commissions (Münch 1971). Turf battles then began at Lake Constance for the institutional ownership of this field of political activity. Municipal politicians attempted to form a cross-border planning group with the name "Euregio Bodensee," which failed because the provincial or state governments (German and Austrian Bundesländer as well as Swiss cantons) countered with the convening of an Internationale Bodenseekonferenz (IBK). The national governments finally established binational spatial planning

Table 4.1
"Waves" of institution building in the Euregio Bodensee and corresponding breakthroughs in motorboat regulation for Lake Constance

1960	International Agreement on the Protection of Lake Constance against Contamination; strong demands for regulating motorboats, start of negotiations
First "wave"	
1970	Council of Europe proclaims the "European Year of Nature Protection" and initiates the first "European Conference of the Ministers of Spatial Planning"
1970	First cross-border networks of environmentalists
1970–1973	First "race" for institutionalizing cross-border planning: local Euregio Bodensee, subnational International Conference on Lake Constance, federal Spatial Planning Commissions
1973	International Agreement on Shipping on Lake Constance: first regulations for motorboats
Second "wave"	
1987	European Community proposes the Single European Market for 1992
1989	European Community introduces a community initiative with financial aid for cross-border collaboration (INTERREG)
1989–1991	Founding of the Lake Constance council, declaring itself "the voice of the people" of the Euregio Bodensee
1990	Lake Constance Conference of Government Leaders enhances its activities and formalizes its institutional basis
1991–1992	New emission standards for motorboats (agreed upon by the Lake Constance Conference of Government Leaders, implemented by the International Shipping Commission for Lake Constance)
1992	Motorboat lobby strengthens its cross-border collaboration
1993	Environmentalists establish more formal cooperation structures

commissions (Blatter 1998). In this discursive and institutional environment, the first breakthrough in cross-border negotiations aimed at regulating the boats on Lake Constance took place in 1973.

These new cross-border institutions now searched for targets for their efforts. Already highly salient, the problem of sport boats seemed to these institutions a suitable field in which to make their mark. After a consolidation phase, the International Lake Constance Conference, as well as the German-Swiss Spatial Planning Conference, extensively exploited the theme of boats on the lake from the end of the 1970s to the present. But only when a new window of opportunity opened at the end of the 1980s and the beginning of the 1990s was the next regulatory breakthrough possible.

In 1982, the Internationale Gewässerschutzkommission für den Bodensee (International Commission for the Protection of Lake Constance), which had previously been concerned with the construction of water treatment plants, issued a report, "Limnological Effects of Shipping on Lake Constance" (IGKB 1982). Highlighting the damage inflicted on the lake by chemical pollutants, this report, in conjunction with the strong publicity generated by the Association of the Lake Constance–Rhine Waterworks, shifted the emphasis of debate from structural damage of the lake's shore zones to the chemical pollution of its water.

Scientific and technical investigations in the 1980s raised hopes for a technical reduction of boat emissions on the lake. But only at the end of the decade, when a new wave of cross-border institution building reached the Lake Constance region, did these investigations produce political results. This time it was not the Council of Europe but the European Community that induced a renewed competition for the institutional ownership of this field by convening of the European Single Market in 1992 and the creation of an assistance program for border regions (INTERREG) in 1990. The IBK was challenged at the beginning of the 1990s by a second, now successful attempt at a grounding of a regional cross-border institution, the Lake Constance Council. The IBK used the opportunity to reshape itself as a successful and innovative cross-border institution, committed to imposing boat emissions standards whose strictness would be unparalleled worldwide.

By conceiving of cross-border coalitions and institutions as primary sources of the thematization of environmental problems, rather than answers to such problems, two issues can be better explained. First, such a conception explains the progression of cross-border regulations in "waves." Regulation occurs in phases in which the theme of cross-border cooperation offers great political rewards, and in which institutional competition requires legitimization of these institutions. For successful cross-border regulation, it is quite important to "ride" the waves of integration discourses and institution building to take major steps toward innovative environmental policy.

Second, an institutional focus also explains why in this particular case regulation of boats concentrated on chemical pollutants. Whereas at the genesis of the cross-border coalition of environmental protectionists, the structural impacts of shipping were accentuated, the problematization of chemical pollution came into the foreground because of the task orientations of the important political-administrative institutions (IGKB, Arbeitsgemeinschaft der Wasserwerke Bodensee-Rhein (AWBR) (Association of the Lake Constance–Rhine Waterworks)). From the same problem, although not a common one (destruction of habitat), a common problem was created (water pollution), as political cross-border institutions searched for fields of activity with high symbolic value and the scientific-administrative institutions surveyed the problem through the lenses of their task-specific glasses.

The Framing of the Problem through Dominant Policy Paradigms

Although the emergence of the boat problem and its definition as a cross-border issue can be explained extensively by institutional factors, one question remains: Why did the political debate concentrate on leisure boats until the middle of the 1980s, whereas the extended public shipping escaped public scrutiny?[5]

Public shipping is responsible for one-third to one-half of both material and structural damage to the lake. Auto ferries and pleasure boats from communal and state shipping associations travel on Lake Constance.[6] The intensity of focus on private motorboats until the 1980s cannot be adequately explained by a "liberal" or "pluralistic" stance, which is one of the modern approaches to international cooperation described in chapter 1. The owners of these boats are predominantly wealthy individuals, well organized into associations and unions. A significantly more plausible explanation is provided by the fact that redistributive policies, which favored working classes over the more affluent, were being widely adopted across Europe at the time.[7] Majone (1993, 97; 1996, 612) finds a general political change from a "redistributive orientation" to an "efficiency orientation" in the 1980s. Whereas redistributive policies dominated in the first three decades following the Second World War, policies more conducive to the creation of wealth and efficient economic development have prevailed since the 1980s and are currently in favor. A preference for redistributive policies can explain why affluent boat owners became a favored target of the party politicians (in particular, the social democrats) in the parliaments of the riparian states during the 1960s and 1970s.

The change diagnosed by Majone is confirmed today (in the 1990s), as the planned entry of motorized catamarans into public shipping on the lake takes center stage in cross-border regulatory discourses. The dominant discourse now focuses on the nondistributional question of better transport versus better environmental conditions rather than on restricting wealthy private motorboaters for the benefit of the masses. Such a postmodern constructivist or culturalist approach can also better explain differences in various national preferences for boat regulation than one that concentrates on the specific material interests of Lake Constance. Although scientific and administrative experts planted the imperiling of Lake Constance in public consciousness, it was primarily political actors who constructed the motorboats as the focus of the threat. In the egalitarian climate of 1968, wealth represented an especially effective point of attack for social democratic politicians. This became the dominant political message on the German and Austrian sides, since their political culture is more overtly marked by class cleavages than is the political culture of Switzerland.

A territorial cleavage amplifies the class struggle in both Germany and Austria. To a large extent in both states, yacht owners come from the national (Vienna, and earlier in this century Berlin) and subnational (Stuttgart, Munich) centers. Not surprisingly, residents of the Lake Constance region resent their perceived domination by those centers. As a result, Lake Constance–area politicians in Germany and Austria found a winning issue in boat regulation. In Switzerland, by contrast, the class struggle has minimal political resonance, and yacht owners are not bigwigs from the political and economic centers (Zurich has its own lake). Most Swiss boaters on Lake Constance are residents of the Lake Constance region.

In summary, this discussion shows that institutions and dominant ideas are centrally important to the framing[8] of a problematic situation, illustrating the contention in chapter 1 that problem constructions and the need for building shared identities often form the impetus for action in cross-boundary water issues. The water of Lake Constance is used here as an identifying symbol for the building of cross-border communities and institutions in times when new policy ideas challenged the predominance of nation-state-focused politics. Without including these factors, one cannot accurately explain when an environmental problem will be placed on the political agenda, how it will be defined, and which position the interested political actors will take.

Beyond Shielded Sectoral Regimes: Institutional Variety and Rivalry

Some theorists who assume that nation-states are rational actors base their assumptions on the belief that the international political arena is much less differentiated and complex than are domestic arenas (Zürn 1992). This is an important basis of the hypothesis that strategic, calculated actions dominate in international relations. From such a perspective, the focus of analysis is upon negotiations among unified states within specialized institutionalized arenas. In contrast to domestic politics, in which the main lines of conflict develop between sectoral interest groups and their political organizations, international environmental regimes are seen as "sectoral legal systems" (Gehring 1991) in which divisions arise between concerned states.

The purpose of this section is to show that international arenas, as exemplified by Lake Constance, are often as differentiated and complex as domestic arenas; therefore, the assumptions of these theorists lead us in the wrong direction. The notion of glocalization suggests that this level of complexity will become more common in the future.

We can differentiate three types of cross-border networks and institutions: transgovernmental commissions, transborder coalitions, and regionalist associations. Table 4.2 provides an overview of the cross-border networks and institutions involved in the regulation of boats on Lake Constance. The following sections will provide additional information about the most important cross-border linkages.

Transgovernmental Commissions

Transgovernmental commissions on Lake Constance include the Internationale Bevollmächtigtenkonferenz für die Bodenseefischerei (International Commission for the Fisheries of Lake Constance), the IGKB, the Internationale Schiffahrtskommission für den Bodensee (International Shipping Commission for Lake Constance), and the Deutsche-Schweizerishche Ramumordnungskommission (German-Swiss Spatial Planning Commission).

The IGKB is the central authority for protection of Lake Constance. In 1967 and 1987, the IGKB passed the "Guidelines for the Preservation of Lake Constance." In the mid-1980s, the IGKB expanded its mission beyond purely material (physical-chemical) investigation (compare the Guidelines from 1967 and 1987). An expert committee, with no connections to business interests, performs the IGKB's work, in which representatives of scientific institutions collaborate with representatives of the water conservation and environmental protection departments of the bordering nations or cantons in the catchment area. Within the IGKB, intense feelings of community have been built. These feelings of community are not only based on a common view of the problem, which is, according to Haas (1989, 1992) characteristic of "epistemic communities;"[9] they are also based on an atmosphere of friendship.[10] Over time, cooperation has taken on an increasingly informal and pragmatic form. An ambassador initially headed the delegation leadership, but a leader of the appropriate governmental department now occupies this position; within the delegation, lawyers were replaced by engineers (interview with Mr. Walter Schnegg, Environmental Protection Agency, Kanton Thurgau, June 8, 1994).

In 1966, the representatives of the riparian states met to reframe the "International Shipping and Harbor Regulations for Lake Constance,"

Table 4.2
Major cross-border institutions at Lake Constance (ordered by type of institution and by relevance for the regulation of motorboats on Lake Constance)

Name	Founding date (major steps of institutionalization)	Goals/tasks
Internationale Gewässer-schutzkommission für den Bodensee (IGKB) (International Commis-sion for the Protection of Lake Constance)	1960 (International Agreement on the Protection of Lake Constance against Contamination)	Water protection, expanded in 1987 to ecosystem protection
Internationale Schiffahrt-skommission für den Bodensee (ISKB) (International Shipping Commission for Lake Constance)	1972 (International Agreement on Shipping on Lake Constance)	Regulation of ships and boats on the lake
Deutsch-Schweizerishche Ramumordnungskom-mission (DSRK) (German-Swiss Spatial Planning Commission)	1973 (without a formal treaty)	Spatial planning trans-portation planning
Internationale Bevoll-mächtigtenkonferenz für die Bodenseefischerei (IBKF)	1893 (Agreement on Implementation of Similar Regulation for the Fisheries on Lake Constance)	Regulating the fishing in-dustry of Lake Constance
Arbeitsgemeinschaft der Wasserwerke Bodensee-Rhein (AWBR) (Asso-ciation of the Lake Constance–Rhine Water-works)	1968 (association based on private law)	Water supply and main-tenance
Umweltrat Bodensee/ Bodenseestiftung (Envi-ronmental Council of Lake Constance/ Foundation for Lake Constance)	1970 (first cross-border networks) 1993–1994 (formal institutionalization)	Spatial planning, trans-portation planning

Membership	Type of cross-border institution
Signatory members: Swiss federal government, cantons St. Gallen, Thurgau; Austrian federal government; German Länder Baden-Württemberg and Bavaria official observer: German federal government	Transgovernmental commission
Informal members: Austrian Land Vorarlberg, Swiss cantons Schaffhausen, Graubünden, Appenzell	
Signatory members: German, Swiss, and Austrian federal governments	Transgovernmental commission
Informal members: subnational agencies from the level of the cantons and Länder	
Signatory members: German, Swiss, and Austrian federal governments	Transgovernmental commission
Informal members: subnational agencies (cantons and Länder administration)	
Signatory members: Swiss Eidgenossenschaft, kingdoms of Württemberg, Baden, Bavaria, Liechtenstein, and Austria-Hungary	Transgovernmental commission
66 waterworks owned mainly by local governments	Transnational coalition
Signatory members: German, Swiss, and Austrian federal governments	Transgovernmental commission
Informal members: subnational agencies (cantons and Länder administration)	

Table 4.2 (continued)

Name	Founding date (major steps of institutionalization)	Goals/tasks
Internationale Bodensee-konferenz (IBK) (Lake Constance Conference of Government Leaders)	1972 (first meeting) 1990 (statute, budget, but no international treaty)	In the 1970s and 1980s, focused on joint planning in the Lake Constance region; since the 1990s, has expanded agenda to all policy fields, especially economic development
Bodenseerat (Lake Constance Council)	1995 (informal network)	Exchange of best practices around the lake, especially in the fields of water, transport, and telecommunication
Arbeitsgemeinschaft Bodensee-UferGemeinden (ARGE BUG) (Association of Riparian Municipalities)	1995 (informal network)	Exchange of best practices around the lake, especially in the fields of water, transport, and telecommunication
Parlamentariertreffen der Bodenseeländer und -kantone (Lake Constance Conference)	1994 (informal meetings)	Exchange of information in all policy fields; "controlling" the executive-dominated cooperation at Lake Constance
Internationaler Bodensee-Motorboot-Verband (IBMV); Internationale Wassersport-gem. Bodensee (IWGB); Arbeitsgemeinschaft Freizeit und Natur (ARGE FUN) (International Lake Constance Motorboat Union; International Water Sport Organization of Lake Constance; Study Group Tourism and Nature)	1970/1982/1992 (associations based on private law)	Lobbying for the interests of the boaters on the lake

Membership	Type of cross-border institution
German Länder Baden-Württemberg and Bavaria, Swiss cantons Thurgau, St. Gallen, Schaffhausen (from the 1990s on Appenzell and Zurich) and the Austrian Land Vorarlberg	Regionalist association
44 delegates from the German (21), Swiss (15), and Austrian (6) sides, plus two from Liechtenstein; delegates are political leaders, as well as people from the business sector and universities	Transgovernmental commission
20 municipalities from around the lake	Regionalist association
Members of Parliament from the IBK Länder and cantons	Regionalist association
IBMV: 32 national organizations as members; IWGB and ARGE FUN: national organizations of boaters and fishermen; sport associations and automobile clubs	Transnational coalition

Table 4.2 (continued)

Name	Founding date (major steps of institutionalization)	Goals/tasks
Internationaler Bodensee-verein (IBV) (International Lake Constance Association)	1908 (association based on private law)	Promoting tourism
Internationaler Verband der Bodenseefisher (International Lake Constance Fishery Group)	Beginning of the 20th century	Lobbying for the interests of the fishermen on the lake
Arbeitsgemeinschaft der Bodensee Industrie- und Handelskammern (Association of the Chambers of Commerce and Industry)	1975 (informal network)	Lobbying for the commercial interests in the Euregio Bodensee

which had been in effect since 1867. After measures against motorboats were intensively debated among the public and within the commission, Germany, Switzerland, and Austria signed the "International Agreement on Shipping on Lake Constance" in 1973. The International Shipping Commission for Lake Constance was simultaneously institutionalized and, on the basis of this agreement, the BSO was implemented and eventually incorporated into the national codes of laws in Germany, Switzerland, and Austria. In contrast to its position in the IGKB, the Federal Republic of Germany is not only an observer, but also acts as the German signatory and delegation leader. Through representatives of the national ministry of transportation, the subnational levels (all states and cantons) are represented in the delegation. There are likewise three national delegations. The Shipping Commission does not have its own bureau or authority (Müller-Schnegg 1994, 116–117). Although only representatives of the state control boards are represented in the commission, the members of this commission generally resist restrictive measures on boats, thereby demonstrating a view that is similar to the view held by boat owners.

Membership	Type of cross-border institution
74 municipal tourism offices, national tourism associations, private travel agencies	Transnational coalition
Private fishermen	Transnational coalition
Chambers of commerce from all three riparian states	Transnational coalition

In summary, successive decades have witnessed a plethora of cross-boundary linkages and alliances. Often these institutions and alliances are sectoral, expressing narrow perspectives on problems. Yet such alliances are often aggressive in attempting to expand their portfolios. When not entirely absent from these alliances, nation-state representatives are usually only minor actors. Within them, municipal and other subnational officials interrelate directly. Often, linkages are founded upon transnational communities of experts not unlike those discussed in chapters 8 and 9.

Transnational Coalitions
Cross-border associations are identified as transnational coalitions if their representatives are social actors distinguishable by shared interests and values. The AWBR, the Unweltrat Bodensee (Environmental Council of Lake Constance), and the various associations of the boating lobby all play important roles in the regulation of boats on Lake Constance. Likewise, but only marginally, the Internationaler Bodenseeverein (International Lake Constance Association) (as a representative of

the tourism industry), the Internationaler Verband der Bodenseefischer (International Lake Constance Fishery Group), and the Arbeitsgemeinschaft der Bodensee Industrie- und Handelskammern (Association of the Chambers of Commerce and Industry) participate in the water conservation arena. Practically all social interest groups on Lake Constance have built cross-border associations, resulting in the establishment of what the authors of chapter 1 refer to as nonterritorial communities based on shared values or interests in which it becomes difficult to distinguish national affiliations.

Initiated by the waterworks on Lake Constance, the AWBR is a proponent of preserving the lake. Its power is based on the critical function of its members (drinking water providers for 10 million people), their political position, and their organizational authority. On June 7, 1968, nine Swiss and nine German waterworks came together in Constance as AWBR "to divert and remove dangers from the public water supply" (translated from Article 2, Paragraph 1, of the statutes of the AWBR). In the following years, communal waterworks from Switzerland, Liechtenstein, Vorarlberg, Bavaria, Baden-Württemberg, and Elsace also joined, boosting the AWBR's membership to sixty-six during the 1990s (Weindel et al. 1993, 520).[11]

Although it has no independent secretary and its budget of 300,000 SFr is almost entirely allocated to its scientific research program, the AWBR, as authorized by both German and Swiss law, can undertake wide-ranging initiatives. An executive committee, a scientific council and topic-specific study groups provide the organizational preconditions that allow the waterworks to coordinate their individual scientific capacities to work toward a common objective. The perceived "political neutrality" of its statements is responsible for AWBR's widespread recognition and legitimacy. Its extensive presence and local anchoring has allowed the organization to engage in massive political mobilization while serving as a font of information. The AWBR is focused primarily on surface water and chemical pollution and has strongly influenced the discussion about boats on Lake Constance to move toward a focus on pollution from hydrocarbons.

When the second wave of institution building swept over the Lake Constance region in the beginning of the 1990s, environmentalists formalized cross-border cooperation, which to this point had only occurred through a loose network. In 1993, eighteen environmental organizations from all three countries founded the Environmental Council of Lake Constance. The founding of a Lake Constance Endowment followed in 1994.

Reacting to the first demands for the restriction of motorboats on Lake Constance in the 1960s, motorboat owners formed the Internationaler Bodensee-Motorboot-Verband (International Lake Constance Motorboat Union). In 1992, another lobby of water sports enthusiasts concerned with the lake was formed: the ARGE FUN (Arbeitsgemeinschaft Freizeit und Natur, or Study Group on Tourism and Nature). Thus the sport associations became tied in with overt lobbying organizations with clear policy agendas. In addition to distinctive publicity work, these groups financed scientific studies that disputed the negative consequences of boating on the lake.

In sum, like the transgovernmental commissions, nongovernmental groups have formed cross-border linkages along sectoral lines based on shared interests and values.

Regionalist Associations

Cooperative forms are labeled regionalist associations if they are not anchored in international law but are supported by elected politicians. Regionalist associations define themselves territorially in regard to a cross-border region and are not sectarian in regard to a certain objective.

When the idea of a "Euroregion" first came to the Lake Constance region at the beginning of the 1970s, only the institutions of the subnational governments (IBK) were able to firmly establish themselves. During the second wave of the 1990s, other regionalist associations were added: the Bodenseerat, the Parliamentary Commission of Lake Constance, and the Association of Riparian Municipalities, all groups with common interests in water conservation and regulation of boats on the lake.

The IBK[12] has developed into a highly visible and active regionalist association with cross-border activities in all major policy fields. Government leaders of the relevant countries and cantons meet twice each year. The IBK is not based on an international treaty; its decrees have no binding power. Since the beginning of the 1990s, a standing committee and various study groups have facilitated cooperative research among many disciplines and specialists. The themes of water conservation and, in particular, the increase in motorboats on the lake have been the focus of IBK activities since its inception. With intensified cooperation since 1979, the water conservation problems have been handled internally, first in the standing committee, and in a subcommission on exhaust fume regulations since 1984 (Müller-Schnegg 1994, 96–106). The IBK played a successful role as mediator in the conflict between environmentalists and boat owners during the 1980s.

After two previous Lake Constance Forums, the inaugural meeting of the Bodenseerat (Lake Constance Council) was held on a ship in the middle of Lake Constance on September 25, 1991. The Lake Constance Council proclaimed itself to be "the voice of the people" of the region. Since that time, it has given priority to the advancement of the regional economy (stimulated by the increasingly competitive struggle within the limits of the Single European Market of 1992), and its environmental study group also tackles the problem of boats on the lake.

The polity idea of a "Euroregion," inspired by the European unification process, has engendered an all-encompassing interweaving of important elements of the political-administrative systems in the Lake Constance region. Social and subnational political actors have institutionalized their cross-border interaction and relegated the nation-states to playing only a marginal role in political decisions. In conclusion, it may be asserted that political regulation in the Lake Constance region developed amid the background of a highly differentiated institutional structure. The following section depicts this structural complexity and its effects on the process of political decision making.

Beyond International Negotiations: Antagonistic Advocacy Coalitions and Regionalist Mediators

This section discusses postmodern cross-border systems of negotiation, in which territorially defined actors like nation-states can no longer be conceptualized as unitary actors with consistent preferences. The first step of a cross-border negotiation is not national preference building,[13] but rather the development and stabilization of transnational "advocacy coalitions." In the case of Lake Constance, transgovernmental commissions have been bound in these coalitions as well and have developed the qualities of independent actors. Political struggles occur primarily between these organized advocacy coalition focal points. Mediation then follows through a territorially defined regionalist association. The "two-level game" (Putnam 1988) in which the "domestic game" or subsystemic level stands distinct from the "international game" or systemic level, is effectively turned upside down.

The Development and Institutional Stabilization of Transnational Advocacy Coalitions

Paul Sabatier developed the "advocacy coalition" framework in order to analyze decision-making processes in the domestic political arena. According to this perspective, antagonistic advocacy-coalitions compete in specific political fields and attempt to influence political decisions according to their divergent conceptions of an issue. Advocacy-coalitions are composed of persons in various positions (elected officials, politicians and administrative officials, presidents of interest groups, scientists) that share a specific belief system and that exhibit coordinated behavior over a long period of time. A "belief system" is defined as a "set of fundamental values, causative hypotheses and problem preceptors" (Sabatier 1993, 127). In addition, so-called policy brokers may step forward to serve in mediating roles. This approach appears particularly well suited to explain the processes of the cross-border regulation of boats on Lake Constance.

The environmentalists and their cross-border organizations, established first in the 1960s, could be described as the first advocacy coalition. The belief that a significant presence of boats and ships on Lake Constance represents a danger to the natural and cultural heritage of the area is the shared normative-cognitive basis of this coalition. The ANU, the AWBR, the Commission on Water Conservation, and the IBKF demanded radical measures against boats on the lake in their position statement concerning a new Lake Constance Shipping Ordinance at the end of the 1960s. A counter-coalition formed as a reaction, composed of the International Lake Constance Motorboat Union and the International Shipping Commission for Lake Constance. These actors share the conviction that the calculated chemical pollution of the lake can be dismissed as statistical artifice. Their argument contends that licensing regulations on the German side artificially inflate the actual number of boats on the lake. (Essentially, the Germans mandate that all boats that lie on land or in other locations must be registered, thus the count is more accurately described as a measure of the potential (as opposed to actual) usage of the lake.) Accordingly, both coalitions subscribe to very different normative-cognitive definitions of the problem. The interests of the boating lobby held sway during the first round of regulation at the beginning of the 1970s, thanks to the transfer of jurisdiction over regulation to the newly formed (and sympathetic) Shipping Commission. The administrative actors in this commission act on the basis of their close relationships to their clientele; that is, they behave less as control boards than as representatives for the interests of the boat owners.

The building of international lobby organizations is quite common in the context of international regimes; the crucial difference in the creation of international coalitions in the case of Lake Constance is that the cross-border regulation of boats takes place in the space between two sectoral regimes.[14] The transgovernmental commissions, the Commission on Water Conservation, and the Shipping Commission were not the object of social interest groups; they were themselves the center of the advocacy coalitions. This became clearer in the following years as the conflict shifted from a public exchange of blows in the regional media to actions behind the scenes, with each faction marshaling scientific studies and designing programs to buttress its respective arguments. (For a more in-depth analysis of the role of media in transboundary water conflicts, see chapter 6.)

A strong rivalry and intense distrust characterized the relationship between the IGKB and the Shipping Commission during the 1970s. According to statements of members of the two commissions, the struggle over the provisions of the Lake Constance Shipping Ordinance was "a fierce fight." For example, each side perceived drafts proposed by the other side as "traps."[15] These statements reveal that the actors perceived the sectoral commissions, and not the territorial political entities, as the primary players in the negotiation process. Differences between the national (or subnational) representatives within the commissions were of only secondary importance.

Although a majority of government leaders clearly supported restrictive measures against boats on Lake Constance under the auspices of the International Lake Constance Conference, a veto by the Swiss canton of Thurgau precluded a decision in 1984. Instead, the IBK convened a subcommission on exhaust fume regulations. Transportation and environmental administrations of states and cantons as well as representatives of the Water Conservation and Shipping Commissions were all represented. Although the German administrations worked hard to establish a scientific-technical basis to justify limits on emissions from boat traffic on the lake, it was primarily the pragmatic and consensus-oriented Swiss who brokered a compromise between the two commissions. A trusting relationship developed between the chairman of the subcommission (who was also director of the shipping office of the canton of St. Gallen) and the Thurgau representative of the water conservation administration during many shared train rides to the capital of Baden-Württemberg, Stuttgart. In this way, the deadlock between the two commissions was eventually broken. But it took another "wave" of Euroregion building until this compromise could finally be transformed into legally binding cross-border regulations.

After the members of the IBK subcommission on exhaust fume regulations agreed on measures, the Shipping Commission once again handled implementation of emissions standards. The IBK, however, did not miss the opportunity to present the emissions standard to the public in one of its meetings and a 1991 report. The government-appointed (not publicly elected) Shipping Commission was forced to allow publicly elected subnational politicians to present themselves as the regulating institution of Lake Constance because the regional politicians held democratic legitimization of their role as representatives for the region. Defending its turf, the Shipping Commission countered with a reference to its professional expertise and reputation. The Shipping Commission pointed out that the recommendations of the IBK subcommission were incomplete and insufficient when received. Such rivalries between the cross-border institutions are just one side of the extensive cooperative structure on Lake Constance. The next section assesses the fortunes of such a differentiated institutional landscape.

Between Territorial and Sectoral Identities/Loyalties: Chances for Multiple Differentiation and Integration Processes

Assessed in retrospect and in view of what it has achieved, the cross-border cooperation on Lake Constance appears to be a success. With the establishment of emissions standards for boats, an incentive now exists for the production of cleaner, quieter motors for boats. Because the newer, cleaner motors and the boats that contain them are more expensive, the possibility that the proliferation of boats on the lake can be stopped also exists. The various forms of cooperation described earlier have been essential preconditions for this success. An impasse on this issue would probably have resulted had there been only one arena of interaction where aggregated national interests were matched against one another. On the Swiss side, a position against restriction and regulation of boats would certainly have been adopted, whereas the Austrian and German sides would have persisted in their radical demands. As consensus was necessary for resolution, crafting a common regulation would have proven difficult, if not impossible.

The case of Lake Constance also demonstrates that functional, cross-border associations, which can overcome territorial positions on the basis of their common view of the problem, cannot unilaterally lead to innovative solutions. As the first round of institution building in the 1970s shows, cross-border advocacy coalitions tend to be narrow focused and what we described in chapter 1 as essentialistic. They could therefore also be implacably opposed to each other and hinder the search for compromises. Therefore, the search for better forms of governance should not be geared toward a purely functionally focused "architecture of governance," as some public choice authors advocate (for example, Frey and Eichenberger 1996; Casella and Frey 1992). Only the differentiation of functional and territorially defined institutions and their processual recombination allow a productive cooperative process and the development of innovative solutions.

Summary and Conclusion

The cross-border cooperation on the regulation of boats on Lake Constance is an instructive example of the kind of environmental politics to be expected with growing frequency in times organized through intermestic politics and multi-level governance. The chapter has shown that the basic motivation for cross-border environmental policy is not the recognition of common-pool problems by actors, but rather the need of institutions to increase their profile and legitimacy. For the politicians at Lake Constance, the central meaning of the lake's water is not so much a natural gift that has to be saved as a valuable symbol for creating identity and legitimacy for political institutions in the Euregio Bodensee. Water quality negotiations are no longer restricted to sectorally bounded institutions nor characterized only by territorially defined interests. Instead, a broad range of increasingly autonomous cross-border networks and institutions crowd the field of transboundary resource policy.

With respect to theoretical approaches toward transboundary water policies, the case study presented in this chapter argues for the influence of ideas and institutions. A rational approach toward explaining policy processes and outcomes, because of the fragmentation of the institutional environment for cross-border water policy, is increasingly inappropriate. The rational approach conceptualizes institutions as structural environments that constrain strategic action. A focus on context and institutions points to a new political landscape where political actors, communities, and organizations are frequently embroiled in dynamic processes of change and reconstruction. The diminution of the hegemonic position of the nation-state increases both the complexity of political structures and the level of institutional competition, simultaneously opening up new opportunities for environmental policy and rendering outcomes more difficult to calculate. Therefore, recognizing the importance of postmodern, nonmaterial, symbolic meanings and issue framing becomes especially important.

Notes

1. For a more detailed and documented version of this article, see Blatter 1994. This article is a theory-oriented reinterpretation of the original study which was aimed toward policy recommendations.

2. The boating lobby could return to the official statements as to the adequacy of water quality by managers of the various waterworks around the lake to mobilize against the supposed pollution of Lake Constance by boat motors (IBMV 92).

3. Compare to the differentiation of "same problem" and "common problem" in Scherer, Blatter, and Hey 1994.

4. The Europarat is not to be confused with the Council of the European Community (now the European Union (EU)). Almost all European countries are gathered in the Europarat, but it scarcely has any authority. The EU, in comparison, was at this time limited to western European countries and has since developed into a very integrated and important political formation.

5. Only the lobby of leisure boat captains pointed again and again in their defense to the damage caused by the ferries and cruise ships of the public shipping companies.

6. Besides the city of Constance, which has six ferries, the national Bahngesellschaften sends out the other ships on Lake Constance: Swiss Bundesbahn (SBB): 14 ships/1 ferry/877,000 travel guests; German Bundesbahn (DB): 7/1/3.3 million; Austrian Bundesbahn (ÖBB): 7/0/460,000 (1992 statistics in Blatter 1994).

7. Peter Hall has developed the conception of "policy paradigms" for the description and explanation of the change from a Keynesian to monetaristic economic policy in England. He states that "policymakers work within a framework of ideas ... which specifies not only goals and instruments, ... but also the very nature of the problems they are addressing" (Hall 1993, 279). Paradigmatic changes distinguish themselves in that not only instruments, but also problem definitions and objectives, change (Blyth 1997, 233).

8. "Framing" refers to interpretations of reality and problem definition in specific situations. It is a concept that stands in the tradition of Erving Goffman (1980) and has been introduced into policy research most prominently by Rein and Schon (1991). It is now widely used in policy research (for example, Benz 1997, 307; Larason Schneider and Ingram 1997, 73; Nullmeier 1993, 180).

9. Haas defines an "epistemic community" as "a specific community of experts sharing a belief in a common set of cause-and-effect relationships as well as common values to which policies governing these relationships will be applied" (Haas 1989, 384).

10. One interview subject said, "The IGKB is the most beautiful international commission there is."

11. In 1970, the AWBR came together with the Association of the Rhine-Waterworks E. V. (ARW), which was started in 1953 and which, with the

Rijncommisie Waterleidingsbedrijven Holland (RIWA) begun in 1951, became on January 23 the International Association of the Waterworks on the Rhine Catchment Area (IAWR). (Schalekamp 1988, 3).

12. The official name has been replaced mostly by the name "International Lake Constance Conference."

13. See for example the statement of one of the most prominent advocates of "liberal intergovernmentalism," which is the dominant modernist approach in international relations: "States first define preferences.... Then they debate, bargain, or fight to particular agreements" (Moravcsik 1997, 544).

14. In contrast to the normal situation in which lobby groups form international coalitions in order to influence intergovernmental negotiations within a single regime, in the case of Lake Constance we find two intergovernmental commissions, together with "their" interest groups, building two distinct international coalitions. The primary conflicts arise not between various (national) positions within one regime but between the two sectoral regimes (shipping and water conservation).

15. These wordings are drawn from interviews with members of the commissions.

References

Benz, Arthur. 1997. "Policies als erklärende Variable in der politischen Theorie." In Arthur Benz and Wolfgang Seibel (eds.), *Theorieentwicklung in der Politikwissenschaft—Eine Zwischenbilanz*, pp. 303–321. Baden-Baden, Germany: Nomos.

Blatter, Joachim. 1994. Erfolgbedingungen grenzüberschreitender Zusammenarbeit in Um-weltschutz. Das Beispiel Gewässerschutz am Bodensee. EURES discussion paper 37. Freiburg.

Blatter, Joachim. 1998. "Entgrenzung der Staatenwelt? Politische Institutionenbildung in grenzüberschreitenden Regionen in Europa und Nordamerika." Ph.D. diss., Martin-Luther-University of Halle-Wittenberg, Germany.

Blyth, Mark M. 1997. "Any More Bright Ideas? The Ideational Turn of Comparative Political Economy." *Comparative Politics* 29, no. 2 (January): 229–250.

Breitmeier, Helmut, Thomas Gehring, Martin List, and Michael Zürn. 1993. "Internationale Umweltregime." In Volker von Prittwitz (ed.), *Umweltpolitik als Modernisierungsprozeß. Politikwissenschaftliche Umweltforschung und -lehre in der Bundesrepublik*, pp. 163–192. Opladen, Germany: Leske and Budrich.

Casella, Alessandra, and Bruno Frey. 1992. "Federalism and Clubs: Towards an Economic Theory of Overlapping Political Jurisdictions." *European Economic Review* 36: 639–646.

Deutsche Umwelthilfe, ed. 1991. Wassersport und Naturschutz am Bodensee. Die Position der Naturschutzverbände zum Wassersport. Radolfzell, Germany.

Frey, Bruno S., and Reiner Eichenberger. 1996. "FOCJ: Competitive Governments for Europe." *International Review of Law and Economics* 16: 315–327.

Gehring, Thomas. 1991. "International Environmental Regimes: Dynamic Sectoral Legal Systems." In *Yearbook of International Environmental Law* (1990), pp. 35–56. London: Graham & Trotman.

Goffman, Erving. 1980. *Rahmen-Analyse: Ein Versuch über die Organisation von Alltagserfahrungen.* Frankfurt am Main: Suhrkamp.

Graf-Schelling, Claudius. 1978. *Die Hoheitsverhältnisse am Bodensee (unter besonderer Berücksichtigung der Schiffahrt).* Zürich: Schulthess Polygraphischer Verlag.

Haas, Peter M. 1989. "Do Regimes Matter? Epistemic Communities and Mediterranean Pollution Control." *International Organisation* 43, no. 3 (Summer): 377–403.

Haas, Peter M. 1992. "Introduction: Epistemic Communities and International Policy Coordination." *International Organisation* 46, no. 1 (Winter): 1–35.

Hall, Peter A. 1993. "Policy Paradigms, Social Learning, and the State." *Comparative Politics*, 25 (April): 275–296.

Internationaler Bodensee-Motorboot-Verband (IBMV). *IBMV-Inside: Newsletter of the International Motorboat Association at Lake Constance.* (April 1992).

Internationale Gewässerschutzkommission für den Bodensee (IGKB), ed. 1982. *Limnologische Auswirkungen der Schiffahrt auf den Bodensee*—Bericht nr. 29.

Internationale Gewässerschutzkommission für den Bodensee (IGKB), ed. 1987. *Zur Bedeutung der Flachwasserzone des Bodensees*—Bericht nr. 35.

Internationale Gewässerschutzkommission für den Bodensee (IGKB), ed. 1994. "Bodenseeschiffsstatistik." Unpublished manuscript.

Landtag von Baden-Württemberg, ed. 1992. "Motorboote auf dem Bodensee. Antrag der Abg. Drexler u. a. CDU und Stellungnahme des Ministeriums für Umwelt." In *Landtagsdrucksache* 11/1110 (December 15), Stuttgart.

Majone, Giandomenico. 1993. "Wann ist Policy-Deliberation wichtig?" In Adrienne Héritier (ed.), *Policy-Analyse; Kritik und Neuorientierung. PVS-Sonderheft 24/1993*, pp. 97–115. Opladen, Germany: Westdeutscher Verlag.

Majone, Giandomenico. 1996. "Public Policy and Administration: Ideas, Interests and Institutions." In Robert E. Goodin and Hans-Dieter Klingemann (eds.), *A New Handbook of Political Science*, pp. 610–627. New York: Oxford University Press.

Manning, Bayless. 1977. "The Congress, The Executive, and Intermestic Affairs: Three Proposals." *Foreign Affairs* 55: 309–315.

Marks, Gary, Liesbet Hooghe, and Kermit Blank. 1996. "European Integration from the 1980s: State Centric v. Multi-Level Governance." *Journal of Common Market Studies* 34, no. 3: 341–378.

Ministerium für Ernährung, Landwirtschaft, Umwelt und Forsten Baden-Württemberg, ed. 1981. *Flachwasserschutz am Bodensee*—Heft 11. Stuttgart.

Moravcsik, Andrew. 1997. "Taking Preferences Seriously: A Liberal Theory of International Politics." *International Organisation* 54, no. 4 (Autumn): 513–553.

Müller-Schnegg, Heinz. 1994. "Grenzüberschreitende Zusammenarbeit in der Bodenseeregion. Bestandsaufnahme und Einschätzung der Verflechtungen politisch-administrativer und organisierter privater Gruppierungen." Ph.D. diss., University of St. Gallen. Hallstadt: Rosch-Buch.

Münch, Walter. 1971. "Pro Euregio Bodensee." *Bodensee-Hefte*, nr. 11 (November): 16–17.

Nullmeier, Frank. 1993. "Wissen und Policy-Forschung." In Adrienne Héritier, (ed.), *Policy-Analyse: Kritik und Neuorientierung* (PVS-Sonderheft 24/1993). Opladen, Germany: Westdeutscher Verlag.

Orbig, Karl-Ernst. 1990. "Gewässerschutz am Bodensee." *Wasserwirtschaft* 80, nr. 7/8: 374–380.

Ostrom, Elinor. 1990. *Governing the Commons: The Evolution of Institutions for Collective Action*. Cambridge: Cambridge University Press.

Putnam, Robert D. 1988. "Diplomacy and Domestic Politics: The Logic of Two-Level Games." *International Organisation* 42, no. 3 (Summer): 427–460.

Rein, Martin, and Donald Schon. 1991. "Frame-Reflective Policy Discourse." In Peter Wagner, Carol H. Weiss, Björn Wittrock, and Hellmut Wollmann (eds.), *Social Sciences and the Modern State: National Experiences and Theoretical Crossroads*, pp. 262–289. Cambridge: Cambridge University Press.

Robertson, Roland. 1995. "Glocalization: Time-Space and Homogeneity-Heterogeneity." In M. Featherstone, S. Lash, and R. Robertson (eds.), *Global Modernities*. London: Sage Publications.

Sabatier, Paul A. 1993. "Advocacy-Koalitionen, Policy-Wandel und Policy-Lernen: Eine Alternative zur Phasenheuristik." In Adrienne Héritier (ed.), *Policy-Analyse: Kritik und Neuorientierung*, pp. 116–148. PVS-Sonderheft 24/1993, Opladen, Germany: Westdeutscher Verlag.

Schalekamp, Maarten. 1988. AWBR-Die Arbeitsgemeinschaft der Wasserwerke Bodensee-Rhein am 7. Juni 1988–20 Jahre jung. (Sonderdruck nr. 1158 aus GWA 1988 des Schweiz, hrsg. vom Verein des Gas- und Wasserfaches), Zurich. pp. 303–321.

Scherer, Roland, and Heinz Mueller. 1994. "Erfolgsbedingungen grenzuber-shcreitender Zusammenarbeit im Umweltschutz: Das Beispiel Bodenseeregion." Discussion paper 34, EURES, Freiburg, Germany.

Scherer, Roland, Joachim Blatter, and Christian Hey. 1994. "Preconditions for Successful Cross-Border Cooperation on Environmental Issues: Historical, Theoretical and Analytical Starting Points." Discussion paper 45, EURES, Freiburg, Germany.

Schneider, Anne Larason, and Helen Ingram. 1997. *Policy Design for Democracy*. Lawrence: University Press of Kansas.

Trapp, Werner. 1994. "Donnerwetter in der Konstanzer Bucht." *Bodensee-Hefte* (July 8): 24–29.

Weindel, Werner, Dietrich Maier, and Max Gutzwiller. 1993. 25 Jahre Arbeitsgemeinschaft Wasserwerke Bodensee-Rhein: 5 Länder—Ein gemeinsames Ziel. In *Sonderdruck aus "GWF—Wasser/Abwasser"* 134, no. 9: 520–525.

Zürn, Michael. 1992. *Interessen und Institutionen in der internationalen Politik: Grundlegung und Anwendungen des situationsstrukturellen Ansatzes.* Opladen, Germany: Leske and Budrich.

5

The Yellowstone to Yukon Conservation Initiative: Reconstructing Boundaries, Biodiversity, and Beliefs

Suzanne Lorton Levesque

This chapter examines the Yellowstone to Yukon Conservation Initiative (Y2Y), a transnational environmental network that is attempting to influence decisions on land use on both sides of the Canada-U.S. border.[1] Y2Y is actively involved in reconstructing beliefs about sustainability and biodiversity in its efforts to reconnect historic wildlife movement corridors, protect watersheds, and set aside sufficient land as habitat for the preservation of a maximum number of species. Y2Y conceptualizes land not as territory, but as habitat for all forms of life. When land is constructed as habitat, its definition expands to include all the elements and ecosystem processes needed to promote biological life. In Y2Y's construction, therefore, land and water are integrally and inextricably interrelated.

The chapter focuses on the critical role of new communication and information technologies in facilitating international networking and the transboundary diffusion of nature-centered discourses. It shows how the Y2Y network makes extensive use of these new technologies to extend the international network and to develop common problem definitions and strategic repertoires designed to shape public opinion and policy.

An Overview of the Yellowstone to Yukon Conservation Initiative

In the early 1800s, the western cordillera of North America was a living tapestry of richly varied landscapes ... of intricate habitats and evolving relationships, created over millennia by an interplay of forces we are only now beginning to understand. Today, in the blink of an eye, the tapestry is unraveling. Forests are cut, rivers dammed ... habitats fragmented [into] islands of extinction. And yet ... the Rocky Mountains of the northern United States and Canada still hold the hope—the best on Earth today—of a fully functional mountain ecosystem, replete with habitat to nurture healthy, viable populations of the grand wilderness icons: grizzly and black bears, gray wolves.... Ours is a vision for the future of the wild heart of North America, the vision of a bright green thread, uncut by political boundaries ... a vital remnant of the once-great tapestry.

The above excerpt was drawn from the Y2Y *Network Handbook* and represent the vision statement of Y2Y. More than 200 participants, including organizations, environmental activists, scientists, academics, government agency and parks personnel, citizen interest groups, individuals, provincial government representatives, foundations, and media representatives from Canada and the United States have become involved in the Y2Y coalition. The 131 organizations currently represented in Y2Y range from small grassroots groups to huge international organizations like The Wilderness Society, the Sierra Club Legal Defense Fund, and the World Wildlife Fund ("October 1998 Network Listing," unpublished coarse-filter communication, October 3, 1998.)

The Y2Y network is coalescing around land, water and other resource-related issues in the region stretching from approximately 150 miles south of Yellowstone Park in the United States to the Mackenzie Mountains in the Yukon and Northwest Territories in northern Canada. Y2Y sees its mission as the creation and maintenance of a life-sustaining series of protected core areas and connecting wildlife movement corridors, insulated from industrial development by transition zones in which human uses are increasingly restricted as their proximity to core areas increases. Existing national, state, and provincial wilderness areas and parks, including Yellowstone, Grand Teton, Waterton Lakes–Glacier, Banff, Jasper, Yoho, Kootenay, and Nahanni, represent the core areas.

Reed Noss, editor of the *Journal of Conservation Biology*, is one of the founders of the science of conservation biology,[2] which informs the reserve model on which Y2Y is based. Noss (1992) argues that the majority of existing reserves and parks created in North America are incomplete ecosystems that are too small to protect viable populations of wide-ranging species. Many of the parks in the Y2Y region were originally designated to protect spectacular landscapes, geologic wonders, or aquatic treasures. The presence of towering peaks, pristine lakes, glaciers, waterfalls, geysers, hot springs, and other natural features governed the selection of these parks. However, the original conceptions of parks and water resources as important for their aesthetic or recreational benefits did not necessitate a consideration of the larger ecosystems that affected the protected areas.

In the evolving social construction of parks, these areas are valuable not primarily for their benefits to humans but for their potential to protect biodiversity[3] and wilderness. In this understanding, the ability of parks to fulfill their role in preserving wildlife diversity may be negatively affected by surrounding development. Lakes and rivers in the parks and throughout the Y2Y region are perceived by network members as outpourings of the basins in which they are formed, and Y2Y participants insist that the ability of these aquatic resources and the parks in which they exist to nurture life is seriously compromised by local or regional land or water use decisions.[4] In Y2Y's perspective, therefore, land and water use decisions in the regions surrounding U.S. and Canadian national parks—the core areas of the linked system of protected areas—must be altered. In addition, new protected areas must be created and critical ecosystems must be restored and connected to complete the envisioned goal (Yellowstone to Yukon Conservation Initiative 1997, 2).

Several trains of scientific thought have been synthesized in the strategic approach promoted by Y2Y. Noss[5] developed the eco-region based reserve design model. Grounded in the sciences of conservation biology, island biogeography and landscape ecology,[6] the model focuses on landscape and environmental characteristics such as elevation, hydrology, soil type, and climate that together determine the types of biological life that may be represented in a specific area. A focus on the characteristics that make up animals' habitats is perceived as essential to ensuring the protection of both terrestrial and aquatic ecosystems. Scientists now argue that even large protected areas or reserves cannot guarantee the survival of various species, especially wide-ranging carnivores, if they are surrounded by development that imperils the life-supporting waters, lands, or biological systems within. Additionally, survival of these species cannot be ensured if barriers to wildlife movement, such as roads, rail lines, fences, or dams, divide protected areas or watercourses. To maintain genetically viable populations, animals must be able to move freely between core areas along natural corridors and fish must be free to migrate to and from spawning grounds.

Because Y2Y focuses on habitat needs in the preservation of species, the effects of land use on water resources—which are critical to all life— are of great importance in the development of the reserve design. This is clearly exemplified in *A Sense of Place*, a compilation of scientific reports Y2Y uses as guidelines: "Increased protection of the Y2Y area would not only benefit large carnivores, but also waters that are key to the production of salmon, trout and other coldwater species. As human populations and exploitation increase, more vigilant protection of these areas will be necessary to protect biodiversity and water quality" (Schindler 1997, 96).

The boundaries of Y2Y have been roughly delineated around the Rocky Mountain Trench, now believed to have been the historic movement corridor for migratory animals that live in the Y2Y region. The steep, rocky ridges of the mountain chains channel migratory animals into the valley bottoms, which are now becoming increasingly inhabited by people and bisected by highways and rail lines (Locke 1997). Y2Y's strategy for preventing further degradation and fragmentation and for restoring historic migratory corridors includes integrating public lands with private lands on which conservation easements have been voluntarily established and confining carefully controlled future development along wildlife movement corridors to transitional zones (Yellowstone to Yukon Conservation Initiative 1997, 2).

Accomplishing these goals will not be easy. The planned reserve system covers 800,000 square miles, cutting through major urban areas and prime mineral, timber, and ranch land (Schneider 1997). As figure 5.1 illustrates, it includes territory in two Canadian provinces (British Columbia and Alberta), in the Yukon and Northwest Territories and in several states in the United States, including Washington, Oregon, Montana, Wyoming and Idaho. Y2Y's success, therefore, depends on eliciting the voluntary adoption of the concept on the part of communities and business, industry, and government leaders along its nearly 2,000-mile length. To obtain the support needed, "Y2Y is dedicated to making reconnections: the human connection to the landscape, ecological connections between land-uses, and local connections to regional and global conservation efforts" (Tabor 1996, 3).[7]

The Yellowstone to Yukon Conservation Initiative held its first conference, "Connections," at Waterton Lakes National Park (Alberta,

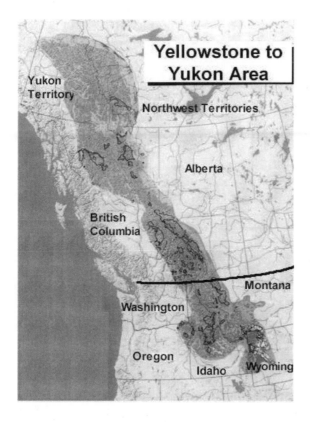

Yellowstone to Yukon Area

☐ **Protected Areas**

Figure 5.1
Yellowstone to Yukon area

Canada) in October 1997. In his opening address, Dave Foreman[8] (1997, 11) told the audience: "We are children of the Pleistocene. Our stories, our myths, our religion, our language, our dreams, our ideas, are all bound up with other large mammals." Foreman's comments and Y2Y's vision statement, excerpted at the outset of this chapter, illustrate the contention in chapter 2 that the impulse to preserve is deeply rooted in a sense of place that depends heavily on symbol, religion, and myth.

Part of Y2Y's goal is to transcend nationalism-based political borders in the creation of its system of linked reserves. In Foreman's words

(1997, 19), Y2Y "is trying to think how ignoring international borders can help us begin to reweave the web of life and wildness."

In the view of Y2Y, integrating the notion of land[9] with "higher" values, such as the preservation of cultural heritage or the "web of life" or the preservation of biodiversity and human security, increases its meaning and social importance and creates a focal point for the creation or strengthening of community and identity. The preservation of land and the waters contained therein then becomes more important than their exploitation. The specific notion of land and water as essential to the preservation of life and community is central to the identity of Y2Y, which adopts as a guiding principle the philosophy expressed by Aldo Leopold (1953, 262) in *A Sand County Almanac*. "A thing is right when it tends to preserve the integrity, stability and beauty of the natural environment. It is wrong when it tends otherwise."

In chapter two, the authors suggest that boundaries around important relationships—whether self, community, or bioregion—are drawn in people's minds. "Ecosystems" or "bioregions" are created through discourse and communication, through collective decisions to include, exclude or emphasize certain relationships.[10] Y2Y is a way of redrawing boundaries and re-imagining territory that both describes the perimeter and ascribes importance to what lies within. The name of the network itself—Yellowstone to Yukon—may be seen as a way of reconnecting to a premodern, natural state of being and meaning, undivided by nation-state imposed borders. In Y2Y's conceptualization, boundaries are not based on lines of longitude, like the 49th parallel that forms the political boundary between Canada and the United States in the Y2Y region, but on watersheds, mountain ranges, and migratory paths, or on "how the landscape lives" (Locke 1997, 35). Y2Y assigns primary importance to biotic processes, to the living landscape—plant, animal and human—that lies within the boundaries of the bioregion surrounding the Rocky Mountain Trench.

Y2Y hopes that "ecosystems and larger scale bioregions may supersede political boundaries in conservation thinking, and provide a framework for integrated land use management"(Tabor 1996, 27). Land and water use decisions must be implemented at the local level; however, Y2Y maintains that land or water use decisions that threaten animal or fish populations or impede their movement in Canada may limit replenishment of genetic diversity and species populations in the United States. Thus Harvey Locke (1997, 45) argues, "This isn't an issue of local impacts, this is an issue of continental impacts." As we will see, this international linkage is an integral and important part of Y2Y's identity and working strategy.

A Two-Pronged Approach

Y2Y advances its challenge to traditional land and water use paradigms in both Canada and the United States along two interrelated dimensions: (1) scientific integrity (use of best-available science) and (2) broadly based activism to build community and activist support for cooperative conservation strategies.

The Y2Y network has devoted enormous amounts of time, energy, and resources to building a strong scientific foundation, which is perceived as essential on two levels. On a strategic level, it offers scientific support for inscribing boundaries around the proposed protected area as well as media credibility and counterarguments to opposition from industry or other interests. On a tactical level, it allows scientists to work in agencies and courts to influence decisions. However, science-based models for conserving biodiversity, including the ones proposed by Y2Y, are admittedly inexact or incomplete. Even in cases of relative scientific certainty, scientific knowledge is often ignored, misused, or dismissed by policymakers. Furthermore, scientific findings alone will have little if any impact on "target populations" without a concomitant change in environmental values. Protecting endangered species and preserving wilderness—and wildness—is fundamentally a sociological and not a biological problem, or as Louisa Wilcox, a leading Y2Y activist, succinctly phrased it: "Science can only tell us how, it can't tell us why."[11] Dr. Jim Butler (1997, 71) used different words to make this same point in his address at the Waterton Lakes "Connections" Conference: "Understanding is not the same. People don't save what they understand; they save what they love."[12]

The second dimension of the Y2Y approach, therefore, relies on facilitating broadly based activism to build support for cooperative conservation strategies within the community. The nature of activist work varies according to the nature of the threats and opportunities within local communities. Public presentations on environmental problems or conservation plans and goals, letter writing campaigns, lobbying of public officials, encouraging consumer boycotts, and providing critical information to community members and leaders exemplify several of the diverse strategies employed by environmental activists at the local level.

Y2Y aggregates and disseminates information, provides a communications network, and facilitates cooperation among its widely separated members at all levels, local to international. Participants contend that Y2Y enhances the credibility of local groups by providing a context for their actions. When viewed within the larger context, "local initiatives no longer appear to be parochial concerns but can be seen as critical keystone efforts" (Tabor 1996, 3). Utilizing this philosophy, Y2Y links autonomous yet interdependent environmental organizations and individuals—through shared ideas—in a joint quest for the preservation of wilderness and biodiversity in the Y2Y region. Y2Y builds the capacity of local and regional grassroots groups, in ways explored in greater depth later in this chapter, to advance the network's principles in their own locales.

Origins and Early Years of the Y2Y Movement

In 1992, The Wildlands Project, a U.S.-based group of conservation biologists and biodiversity activists, published *Wild Earth: Special Issue*, which set forth a plan for the conservation of biodiversity based on the design on which Y2Y has been grounded. Harvey Locke, an environmental attorney in Calgary, Alberta, and at that time president of the Canadian Parks and Wilderness Society (CPAWS) read *Wild Earth* and became convinced that The Wildlands Project model offered the best approach to the protection of the wealth of biodiversity in the Rocky Mountains. Locke called a meeting of various scientists, researchers, and activists involved in biodiversity preservation–related issues to discuss the designation of a system of linked reserves in the area now known as Y2Y[13] (Harvey Locke, personal communication, January 31, 1997).

The following four years were spent assessing and building support for the concept on both sides of the border.[14] During this time, sporadic meetings were held in which strategy, organizational structure, fiscal accountability and fundraising details, among other matters, were hammered out. Various subcommittees were formed to begin or continue tasks such as creating a secretariat and hiring a coordinator,[15] drafting a mission statement and planning a public launch of the project. An assessment of the region's biological status, its ecological threats and its opportunities for preservation efforts was begun; communication linkages among scientists were created and contracts with scientists to provide information on the region were entered into. The reserve design continued to evolve. Work was begun to compile data and establish an electonic communications network through the University of Calgary's Crown of the Continent Electronic Data Atlas.

Y2Y went public at the "Connections" conference. More than 300 people from the region and around the globe attended the conference, which received extensive media coverage in both Canada and the United States. Delegates included representatives from the conservation community, the scientific community, and First Nations as well as Native Americans, academics, municipal and provincial politicians, educators, and private individuals. Local groups were encouraged to provide information on problems, threats, conservation goals, or strategies specific to their locales for use by other groups or the larger Y2Y network. Network communication connections were extended to facilitate the development of a regional strategy in which issues and threats were prioritized, common areas of interest were mapped out, and local actions to be undertaken and barriers to effective local action were identified. The overall purpose was to determine how Y2Y could help support local and regional efforts that would, in turn, support the overarching vision of Y2Y (Legault et al. 1997, 79–95).

Extending and Promoting the Vision

Communication technologies—and the media—have been of paramount importance in the emergence and growth of Y2Y. They have facilitated network building and cooperation among participants and have helped to introduce the Y2Y concept for wilderness and biodiversity preservation to a wider audience. The network often makes very creative use of numerous media venues. Y2Y has its own Internet homepage (http://www.rockies.ca/y2y/) and is prominently featured on the homepages of many of its constituent groups. Y2Y has also been featured on National Public Radio and in the press. Y2Y's message has sometimes been spread in unique and innovative ways. For example, the Y2Y concept was recently featured in a "Wyoming Wear" clothing catalogue. During the spring, summer, and fall of 1998 and 1999, network member Karsten Heuer, a warden in Banff National Park and a wildlife biologist, hiked the entire length of Y2Y. With preplanned stops in numerous communities along the way, this "Walk for Wildlands" introduced the Y2Y concept to schoolchildren and the public and answered questions and public concerns about the Y2Y project. While on the trail, Heuer and his occasional hiking companions "roving" e-mail contact with the outside world, and an Internet site was created for those wishing to check on the progress of the hike.[16]

Brysk (1994, 36) finds that "image bearers" [cultural icons such as the grizzly bear or the wolf in the case of Y2Y] may be used in various media formats to represent complex issues. Posters, coffee table books, brochures, or mailings produced by environmental groups, for example, often contain powerful images—whether beautiful, heart-warming or tragic—meant to evoke strong emotions. Jeremy Wilson (1992) discusses the use of images to foster emotional attachment and build support for preserving a place many of those being asked to offer support will never see. In Canada, according to Wilson (1992, 121), "The slide show has been the bread-and-butter means of delivering the wilderness message." Locke, an eloquent and charismatic spokesperson for the movement, presents a beautifully photographed and powerfully narrated slide show on the Y2Y region and vision. The production, "Reweaving the Wild: Nature from Yellowstone to the Yukon," is both informative and evocative and has been presented to a wide variety of audiences in numerous communities, including an April 1998 showing in Boston to researchers and scholars in the prestigious East Coast academic community. In these ways, and in ways explored in later sections, the network has grown.

The Construction of New Identities and Nature-Centered Discourses

A common feature of these methods of shaping public opinion and extending the network is the prominent use of problem definitions and solutions based on nature-centered discourses. Through the formation of alliances and the diffusion of information, problem definitions, and strategic frameworks, the Y2Y network is actively involved in the creation or evolution of new environmental paradigms and of new identities for its adherents and the human and natural communities in which they live.

Human beings have a need for aspiration and identity in a world that is, for many, becoming increasingly impersonal and confusing. Castells (1996, 354–355) argues that a "crisis of legitimacy" is voiding the meaning and function of industrial-era institutions such as states, labor movements, churches, and the patriarchal family, draining sources of traditional "legitimizing identities." This crisis of legitimacy, according to Castells, is resulting in the emergence of powerful resistance identities, such as social movements, that defend spaces and places on behalf of both humans and nature, claim historic memory, and affirm the permanence of values.

Many Canadian Parks, such as Banff National Park, are perceived by network members as becoming increasingly commercialized. Development pressure around (and even within) many parks in the Y2Y region is also increasing.[17] For example, in the Bow River Valley, which represents approximately 50% of all usable large-carnivore habitat in Banff National Park, development has reduced habitat to 10% of its original extent (Davis 1995). Researchers have recently acknowledged that even Yellowstone Park, despite its enormous size, may not suffice for the long-term protection of some species because of development and other human-caused threats around its borders. Rapid regional development in response to a combination of increased in-migration and decreased out-migration is leading to greater impacts on land, waters, and other resources (Rasker and Alexander 1997). The combination of these and similar factors has led to a heightened sense of encroachment on "sacred places." Jim Morrison[18] (1995, 20) asserts, "Most of the anger and frustration voiced by conservationists is related to the diminishing supply of wild places throughout Canada." An increased public awareness, often as a result of exposés by environmental activists, of the damage to public lands and waters and the depletion of resources from extractive industry processes have added to the sense of encroachment and resulted in a drive to protect the remaining wilderness.

Nature-centered discourses emerge as a response to the ecological crises facing different places (Lipschutz 1996). These discourses act as a focal point for the creation of a community that is not bounded territorially. Slater (1997, 1) asserts that "archipelagos of resistance or reverse discourses that have the potential to be connected across space, but which are also distinct, specific and embedded in local and regional contexts, have emerged in many different societies." The Y2Y network is coalescing around shared values, beliefs, and understandings and shared discourses centered on nature. Y2Y is connected to place in that it has definite physical parameters, yet it also transcends place. As we will see in subsequent sections of this chapter, the use of cyberspace-based networking allows Y2Y to transcend time as well by conferring the ability to elicit immediate response or action from participants residing even in distant locations.

In chapter 2, the authors maintained that social actors create, reconstruct, and imagine place and that place is created through ideas and discourse. Identities are built by shared visions: common beliefs that act as focal points that bind actors together. Advocacy networks like Y2Y provide a sense of community and identity based on the *idea* of place. Many adherents to the Y2Y vision have never and may never set foot on Y2Y soil, yet they identify strongly with the landscape it represents. As David Johns (1997, 60) puts it, "There is no doubt in my mind that the people in this room love the land as fiercely as a mother grizzly loves her cubs. And are prepared to fight for it just as fiercely as a mother grizzly." Membership in Y2Y engenders a sense of belonging to a community working to promote noble, principled goals, including the restoration of the landscape and the natural order. It is therefore a way of reconnecting with something nontransient: Land is steadfast; like water, the web of life flows, changes, yet goes on. Y2Y provides a point of conceptual anchorage, a way of resisting destruction, fragmentation, and change in landscapes—and in lives.

The Importance of Information and Alliances

The following section, like others in this volume, highlights the absolutely essential nature of information and contemporary information and communication technologies to successful cooperation or collaboration on transboundary issues. Information—and its aggregation, distillation, interpretation, and dissemination—promotes the emergence, growth, and potential success of environmental advocacy networks. To be most useful to participants, information must be accurate, timely, and complete; it must also be available and comprehensible to all participants, many of whom may be scattered in geographically distant locales. However, numerous barriers to the efficacious accessing or use of information exist. Joseph DiMento's study of the Black Sea Environmental Program in chapter 9 illustrates that gaps in knowledge or barriers to information flows may seriously impede cooperative efforts. Essential data for transboundary environmental cooperation may be difficult to obtain or unavailable for many reasons: They may be incomplete or outdated, highly technical in nature, or dispersed throughout remote locations or institutions.

For transboundary environmental networks to develop an effective command of information, they must often form alliances through which knowledge is sought, gathered, interpreted, and diffused. A communication infrastructure capable of continuously linking all interested parties must exist or be created and maintained. Network members must have the ability to compile and interpret information, define problems, and develop and promote solutions that further desired goals. In addition, the relative powerlessness of environmental actors in relation to well-entrenched economic interests in the Y2Y region leads to the use of networking and the mobilization of populations to equalize power discrepancies. The following section discusses Y2Y's extensive use of information, information technologies, and alliances to foster the growth, increase the power, and promote the goals of the network.

The Role of Information, Lifeblood of the Network
The provision and strategic application of various types of information binds and legitimizes networks and empowers local groups (Keck and

Sikkink 1998). The information made available to member groups through Y2Y's communication system is derived from a wide array of sources and takes many forms. Among the myriad types of information provided are references to recently published scientific or economic reports or journal articles; texts of relevant newspaper articles or books; accounts of strategic attempts, successes, or setbacks; exposures of alleged violations of environmental laws or standards; and notices of government policy proposals or agency decisions. Grassroots groups make use of the knowledge base provided by their connections with Y2Y to assist them in the design of local agendas, strategies, and campaigns and in promoting the larger objectives of Y2Y.

Y2Y therefore serves as an information clearinghouse, compiling and disseminating the kinds of knowledge needed to build a coalition of groups that might otherwise remain unconnected. Through the auspices of the Crown of the Continent Electronic Data Atlas (CCEDA), Y2Y links its participants in a dense web of information that flows in multiple directions: from periphery to core, core to periphery, node to node. Neither space nor time now separates the periphery from the core or node from node; calls for action in one locality can be answered almost immediately by respondents who are geographically distant. Ideas, ideologies, and strategies diffuse through populations far more rapidly today than they could have a century or even a decade ago. Y2Y thus enables its participants to cohere in a community unbounded by space or time. The linkages formed encourage even geographically remote participants to identify with the Y2Y community and to maintain ongoing involvement, making them feel that they are part of the action.

The Communication Infrastructure   If information is the lifeblood of the network, the ability to aggregate, coordinate and distribute it is the heart of the system. The CCEDA,[19] a joint project of CPAWS, the University of Calgary, the University of Montana, and Waterton Lakes and Glacier National Parks, serves as the information hub for the Y2Y network, providing Internet, list serve, and e-mail linkages among participants.

The CCEDA was established to fill in gaps in existing information structures that make sound ecological planning difficult and costly and to compile and integrate existing data, now scattered among offices of industry, consultants, and governments, in a single database (Craig Stewart, CCEDA Project Coordinator, personal communication, November 20, 1997). Through the CCEDA, Y2Y provides bridges among its participants to provide forums for discussion, debate, and cooperation in which shared beliefs, goals, and strategies are hammered out. The list serve is extensively used for rapid exchange of information, updates on training or funding opportunities, dissemination of media or research reports and requests for assistance. For example, a request for a source of information "suitable for use in court" was sent out on the list serve on April 28, 1998. Within a day, several respondents directed the initiator of the request to the appropriate source (unpublished coarse-filter correspondence to Yellowstone to Yukon Council, April 28–29, 1998). A separate list serve for "action alerts" or requests for immediate assistance was added to the communication network in June 1998. In addition to providing these facilities for centralized participation, Y2Y publishes a roster of telephone numbers and email addresses to facilitate direct communication between or among groups and individuals.

The Need for Alliances: Acquiring Information, Increasing Legitimacy and Influence, Building Consensus and Support
Both Canada and the United States have promulgated statutes that ensure the public's right to obtain information; however, critical data may still be quite difficult to acquire from convenient or orthodox public sources, and acquisition of such information can be prohibitively costly in terms of both money and time. For these reasons, it is often far more efficient to rely on alternative sources of information whenever possible. Through the formation of alliances, information can be acquired and dispersed much more efficiently. The fact that networks are most politically effective if they can claim to represent a truly global constituency (Smith 1997) increases the importance of cultivating ties with other local, regional, national, or international environmental organizations. This section explores several of the various types of alliances that have been or are being formed within the Y2Y network.

The need for local knowledge to assist in the framing of issues and to increase support for the Y2Y concept necessitates linkages with grassroots groups. The linkages that have been formed by Y2Y allow for low-cost monitoring and information acquisition on the part of both the larger Y2Y organization and the grassroots organizations themselves, as information is often exchanged between or among local groups. The ability to gather and disseminate local information over large distances also allows the network to further its goals by bringing hidden instances of local environmental destruction to the attention of a wide audience and by calling for action on specific issues on the part of all members. Forming linkages with local grassroots organizations may also help increase the credibility of Y2Y in the local communities and within the larger region. Because these connections provide information on issues within local communities, Y2Y members are able to speak knowledgeably and sensitively about local and regional problems or concerns as well as about how Y2Y might serve the interests of local communities.

Because isolated local groups with small memberships and few resources are seldom influential (Potter 1995), alliances with larger networks offer numerous advantages to local grassroots organizations. Y2Y directly provides, or provides information about, workshops or informational materials that emphasize the transfer of essential activist and networking skills among Y2Y members. Training workshops or materials previously offered have included (to cite only a few) grassroots development, coalition building, fund-raising, consensus building and conflict resolution, conservation plan development, media and communications strategies, public education, and encouraging youth involvement. The network conducted a workshop dealing exclusively with aquatics issues during the summer of 1999, illustrating the increasing centrality of water resources for reserve design in Y2Y.

Local groups, therefore, receive a wealth of information and resources as a result of their connections to the larger network. Ready access to information on the part of diverse and geographically distant actors can, however, occasionally prove problematic. Grassroots actors may disseminate "facts" in the name of Y2Y with which other members disagree or acquire and disseminate information from sources that others regard as ethically inappropriate, such as foundations established by extractive industries. Disagreements have also arisen over the relative importance and appropriate timing or use of certain information (e.g., scientific or economic findings) in the framing of issues. In times of dissension, the role of the Y2Y coordinator assumes singular importance in the formulation of coherent discourses or strategies. Y2Y currently benefits from the services of Bart Robinson,[20] an affable, diligent, and resourceful coordinator with a gift for elucidating and synthesizing disparate positions and fostering accord among members.

Alliances with national or international environmental nongovernmental organizations (ENGOs) and funding organizations both enhance the credibility of the network and facilitate the aggregation of information by Y2Y. For example, reports published in 1997 and 1998 that resulted from two critically important information-gathering initiatives were funded by large national or international groups or foundations affiliated with Y2Y. The first report, *The New Challenge: People, Commerce and the Environment in the Yellowstone to Yukon Region*, was prepared by Ray Rasker and Ben Alexander (1997) of The Sonoran Institute in Bozeman, Montana, and was funded by The Wilderness Society. It represented the first attempt to examine the economy of the entire Y2Y region and present the information in a comprehensible manner to enable communities to make informed decisions about economic and environmental choices. The second report, *A Sense of Place: Issues, Attitudes and Resources in the Yellowstone to Yukon Region* (cited earlier in the chapter) is the end product of the work of many collaborators and numerous funding organizations, including the Bullitt Foundation, the Foundation for Ecology and Development, the Henry P. Kendall Foundation, the Lazar Foundation, the Newland Foundation and the Wilberforce Foundation.[21] Findings on the hydrological, physical, and biological characteristics of the region, on human influences and attitudes, on conservation efforts and approaches, and on various other trends and issues facing the region are included.[22] These findings will be critical to the development of successful strategies, conservation models, and discursive approaches structured around sustainable development and biodiversity preservation in the Y2Y region.

Alliances with aboriginal cultures assume enormous importance in Y2Y and may indeed be essential to the success of the initiative. According to Morrison (1995, 24) a "new level of authority—Native self-government—is being created in Canada." Virtually all of British

Columbia and large portions of other provinces as well are subject to aboriginal land claims. This obviously has enormous implications for the preservation of wilderness.

As David McDermott Hughes's chapter on the Chimanimani Reserve illustrates, indigenous people have sometimes borne the costs of efforts to protect natural areas. In Canada, native leaders allege that governments have either ignored or abrogated their aboriginal or treaty rights in the course of the last century—occasionally at the behest of conservationists. Conservationists, on the other hand, maintain that without their intervention, damage from development would have been much worse (Morrison 1995).

Common ground does exist between environmentalists and First Nations in the Canadian provinces, as exemplified by numerous recent cooperative efforts among conservationists, First Nations, and government agencies at various levels of the land claim process (Morrison 1995). Moreover, aboriginal groups in Canada have themselves reached out to the international environmental community for their assistance in pressuring government decision makers (Barker and Soyez 1994). As will be demonstrated later in this chapter, alliances with First Nations have resulted in tremendous gains for Y2Y as well as for the aboriginal groups.

In the case of Y2Y, alliances with First Nations have attained critical importance, since environmental organizations in the Yukon Territories are currently excluded from discussions on protected areas. Almost all protected areas in the Yukon have been established through aboriginal land claims (Peepre 1995). These connections with First Nations, offer numerous additional benefits. Prevailing ideas about indigenous cultures embody visions of continuity and connectedness and of respect for and harmony with nature, and represent critical elements in the type of discourse utilized by Y2Y. Linkages with these groups provide a way of reaffirming both the connection with and the importance of these values, which are central to the Y2Y mission. They extend the network and enhance its moral legitimacy and credibility by drawing on the global mandate for the preservation of aboriginal communities and the promotion of human rights. They offer the opportunity, through collaboration on joint efforts, both to develop a better understanding of the perspectives and needs of local indigenous communities and to foster an understanding of the goals of Y2Y within those communities.

Y2Y has benefited greatly from the interest and guidance of various members of the academic community. Tim Clark of Yale University has authored or coauthored numerous books and articles designed to strengthen and facilitate the potential of environmental networks in general to work toward the preservation of biodiversity that have been of great use to the Y2Y network. (See, for example, Clark 1993, 1996, 1997; Clark and Wallace 1998; and Clark et al. 1992.) In addition, Clark and David L. Gaillard coauthored "Organizing an Effective Partnership for the Yellowstone to Yukon Conservation Initiative" (n.d.) to guide the network in specific ways to increase its overall efficacy.

Gary Tabor has also worked closely with the network and has authored *Yellowstone-to-Yukon: Canadian Conservation Efforts and a Continental Landscape/Biodiversity Strategy*[23] for the benefit of Y2Y. Tabor is currently codesigning a study on the linkages between human, animal, and environmental health in the Y2Y region. The study is a joint project among ecologists, veterinarians, and physicians from Wildlife Preservation Trust International, Tufts University, and Harvard University. Researchers will work closely with Y2Y, using maps and information provided by the network to assist them in their endeavor. One of the study's primary goals is to develop new ways of conceptualizing the preservation of biodiversity with an emphasis on the indestructible connection between human and environmental health (Gary Tabor, personal communication, April 25, 1998).

Linkages with prestigious universities and well-respected scholars also help to build the network's credibility within political and social structures. Y2Y's alliance with the academic community has therefore shaped the development of the network, increased its legitimacy, and molded its rhetorical and political strategic approaches as well.

The role of the scientific community in Y2Y is discussed in greater depth later in the chapter, but it is important to note here that alliances with this community are among the most critical of those Y2Y has formed. Shared understandings and linkages between conservation biologists and biodiversity activists in the Y2Y region represent what Peter Haas (1992) identified as an "epistemic community." Epistemic communities, according to Haas, foster the adoption and integration of new bodies of knowledge into belief systems and social practices, changing ideational frameworks (Lipschutz 1996). Harvey Locke and other activists utilized these linkages at the inception and in the development of Y2Y.

Scientific information is of ever increasing importance in the creation of policy measures (Blowers and Leroy 1995). The scientific and environmental communities have now joined forces in shaping the political agenda in the environmental realm. By deploying counterexpertise, the environmental movement has been successful in challenging the monopoly on expertise formerly held by state and corporate interests (Blowers and Leroy 1995). In terms of information, linkages with the scientific community provide Y2Y activists with scientific justifications for their proposals. Importantly, scientific knowledge allows Y2Y to frame the debate in a manner that is based in science and yet compelling to the public it wishes to mobilize.

Using Information: The Importance of Issue Framing

Providing information to support the goals of the network is not sufficient, on its own, to ensure the adoption of these goals within the larger community. Movement goals must resonate, or be consistent with the values, beliefs, and behaviors of the populations to be mobilized. Often, fundamental changes in beliefs or worldviews are needed to embed changed practices in societal institutions, and explanatory frameworks must be developed to account for the necessity of change (Lipschutz 1996). Y2Y reframes issues to make them resonate with the public and disseminates these new frames of meaning to shape public opinion and generate consensus for its goals. Public consensus can then be used to pressure political actors. To elicit support or action, Y2Y often frames issues simply, in terms of right and wrong, resulting in the following type of argument: "The message is clear. Either we manage the northern Rocky Mountains as an integral ecosystem that ensures the long-term survival of wildlife or we watch wide-ranging animals go extinct" (Locke 1996, 30).

Networks reframe issues in the search for hospitable venues in which to promote their goals (Keck and Sikkink 1998). Identities may be transformed in response to opportunity structures and local problems may be reframed in regional or transnational terms to make preferred solutions congruent with the goals of the network. In the Y2Y network, issues have been intentionally framed as an international problem: "Conservation should parallel ecological processes and the movements of species across the 49th parallel" (Harvey 1998, v). A June 9, 1998, list serve posting of a press release entitled "Local Forest Plan Could Destroy Another International Fishery" states: "The Wigwam River is not only the single most important known bull trout spawning stream in the Kootenay Region [in Canada], but it is also of international significance."

Transnational networking offers numerous advantages to Y2Y. At the "Connections" conference, participants listed several of the benefits of transborder cooperation. Among them, "expanding and building partnerships" and "using the U.S. media to draw the attention of Canadians to environmental destruction in Canada" are considered essential. In addition, "Local issues are often best fought when they become national or international issues—in some circumstances an authoritative outsider talking about issues can be more powerful than a local voice" (Minutes of the Nakoda Lodge meeting, December 1–3, 1997).

Networks may increase their effectiveness by framing issues in a way that heightens a sense of urgency (Smith et al. 1997). Y2Y integrates the perspective of Tim Clark (1993, 497), who writes: "Living professional conservationists are the last generation that can prevent the extinction of large numbers of species and the disruption of large scale ecosystem processes." Y2Y echoes the perceived need for immediate action: "Because the Yellowstone to Yukon region is still one of the most biologically intact parts of North America, it makes sense to protect the region now, before more is lost" (Yellowstone to Yukon Conservation Initiative 1997, 9).

In an April 3, 1998, posting to the list serve regarding a cross-Canada "Waterwalk," organized to mobilize action on aquatics issues, the author writes: "If we do not speak up right now for sustainability; make a stand today, at this very moment for something that we believe in— something so simple as pure water—we shall lose it forever" (unpublished communication, Y2Y list serve, April 3, 1998).

Using Information: A "Politics of Purpose"
Brysk (forthcoming) describes a "politics of purpose" in which network actors reframe ideas already present, shift the boundaries of community in existing relationships, introduce new ideas, and contest traditional paradigms. When Y2Y began, its explicit goal was to protect wildlife and maintain connectivity (Harvey Locke, public presentation, Ramada Inn, Helena, Montana, April 24, 1998). The movement now also incorporates human security into its discourse, promoting the idea that healthy human communities cannot be sustained without a healthy biosphere. Incorporating human security, therefore, reframes environmental issues to make them more salient and redraws the boundaries of the community to include those who might otherwise perceive themselves as less concerned with environmental preservation and more concerned with the traditional "maximizing sustainable yield" paradigm. Y2Y also contests traditional economic paradigms: "In a sustainable world, the economy can experience development without growth" (Daly 1987, as cited in Rasker and Alexander 1997, 7). Lastly, Y2Y contests paradigms that relegate the conservation of biodiversity to the Third World: "Y2Y ... demands nothing more of Canadians and Americans than we ask of Kenyans and Tanzanians to protect the great animal populations of the Serengeti" (Locke 1996, 30).[24]

Thus far, we have seen that transnational advocacy networks identify and seek out valuable kinds of information and testimony, reframe issues in a way that makes them more salient, and disseminate these new discursive meanings through various media, using language that dramatizes issues. Information is used to mobilize the public, shape public opinion, generate consensus, and create hospitable venues for environmental preservation proposals. Networks also use their command of information to pursue information, symbolic, leverage, and accountability politics (Keck and Sikkink 1998). The targets of these strategic uses of information may be policy- or decision makers, corporations or other economic actors, the public—or group members themselves. The following section examines various ways in which Y2Y engages in each of these types of politics.

Information Politics   Groups use information politics to link movement goals to preexisting mandates or bases of legitimacy within the international system (Brysk 1994). Y2Y links its goals to the preservation of biodiversity, recognized as an international objective in forums such as the Brundtland Commission or the Rio Conference. Networks also promote change by uncovering and investigating problems and reporting accurate, reliable, timely, dramatic, and well-documented facts (Keck and Sikkink 1998). On May 18, 1998, a "run down of wolves shot in the Rockies and Arizona" was distributed on the Y2Y list serve to the entire Y2Y Council. A few excerpts: "April '95—#10 is ILLEGALLY killed by Chad McKittrick near Red Lodge, Montana. Three years later, judged guilty, sentenced, and appeal rejected, he's still not paid a penny or served a day." "March '98—#39F, the only white wolf in Idaho or Yellowstone, and possibly pregnant, is ILLEGALLY killed east of the Park. Two months later, charges are still 'pending.' "

Y2Y also uses information to attempt to resolve conflicting perspectives among its adherents. Network constituents have been mobilized from locales far distant from one another, in which very different environmental, social, economic, and political conditions prevail and values, goals, priorities, problem definitions, and solution or strategy preferences may vary considerably among participants. Yet the solutions and strategies proposed to further network goals must respond to the underlying values and goals of its membership to elicit broad-based support. Although it does not represent the ideal platform for disputation, the list serve is commonly used as a locus for debate on contested issues. Through the sharing of individual perspectives on issues and through the provision of information to support those perspectives, opinions and attitudes may be modified and consensus may be reached. In this manner the network has developed acceptable issue frames and strategic approaches.

Symbolic Politics Networks promote their own growth by framing issues through identifying and providing convincing explanations for powerful symbolic events (Keck and Sikkink 1998). Issue framing represents "conscious strategic efforts by groups of people to fashion shared understandings of the world and of themselves that legitimate and motivate collective action" (McAdam et al. 1996, 6). These "mobilizing frames" must be crafted out of preexisting cultural materials and must resonate with cultural understandings and goals (McCarthy 1997).

Y2Y makes heavy use of imagery and symbolism in framing issues and crafting mobilizing frames. The following statement was extracted from a promotional brochure published by Y2Y: "These mountains are our temples, our sanctuaries, and our resting places. They are a place of hope, a place of vision, a place of refuge (Yellowstone to Yukon Conservation Initiative, nd)." As mentioned earlier in this chapter, even the name of the network—Yellowstone to Yukon—represents both a symbolic challenge to the rights of nation-states to impose arbitrary and artificial boundaries and a symbolic reintegration of past and present.

An additional example illustrates the use of both information and symbolic politics to target decision makers. During discussions with government officials and in reference to a major highway in Alberta's Bow Valley that is responsible for high rates of animal fatalities, Harvey Locke stated: "When I called it the Berlin Wall of Biodiversity, funny things happened. The Government found some money—$5M—to build wildlife overpasses" (Legault et al. 1997, 45).

Groups pursue symbolic politics when they juxtapose events or information to increase their salience, or when they link facts with testimony (clear, powerful messages that appeal to shared principles) to make the need for action more real to network members or to the public (Keck and Sikkink 1998). As previously mentioned, image bearers (e.g., Yellowstone's icon, the grizzly bear) may be used to represent complex issues. Y2Y's vision statement, excerpted at the beginning of this chapter, provides a perfect illustration of the use of these forms of symbolic politics, as does the following story told by Y2Y member Craig Stewart:

Last week, I joined several wardens and tourists watching the first grizzly of the year as he rolled in the grass and playfully flipped a leg bone (from a white-tailed deer) end over end. With a pair of bald eagles circling overhead against a Vimy Ridge illuminated brightly before a darkening sky, it was quite a sight. Yesterday, that bear was shot, legally, on an adjacent ranch ... this bear was an endangered species six months ago and today he is a rug. The decision to raise the limit and help kill that bear was clearly a political one ... it cannot be justified, biologically. And given American efforts to conserve grizzlies just 8 miles away, as the crow flies, it should not be justified politically either. (unpublished communication, Y2Y list serve, April 9, 1998)

A combination of information and symbolic politics was recently used to great advantage in the Muskwa-Kechika region in British Columbia.

The Muskwa-Kechika has been called the "Serengeti of the North," because it contains the largest combined abundance and diversity of large mammals anywhere outside the Serengeti Plains in Africa. Activists from Y2Y member groups in the region assembled an "impossible coalition" of First Nations, resident hunters, guide-outfitters, trappers, and environmental organizations, many of whom have traditionally been in conflict with one another, to cooperate in achieving the goal of "sustaining the land and wildlife forever." With the strength of this coalition behind them, Y2Y activists managed to build understanding of preservation issues among large numbers of people in the area and persuade them to support preservation goals.

Network members also provided information to politicians to garner their support and became involved in the Land and Resource Management Planning (LRMP) endeavor, a consensus-based process that also included representatives from government and industry. The force of the coalition assembled by the activists encouraged scientists inside the government to use the scientific information previously gathered to support preservation of the region. The combination of these activist strategies eventually resulted in the protection under law of approximately eleven million acres, or nearly one-fifth of the land preservation goals of Y2Y.[25]

Leverage Politics   The goal of leverage politics is to change the policies of powerful actors such as governments or corporations (Keck and Sikkink 1998). Y2Y focuses the attention of its members, and through them that of the general public, on specific environmental issues or concerns to generate transnational constituencies for its environmental goals, thereby mobilizing populations that can exert pressure on national governments.

Moral leverage, or the "mobilization of shame," involves holding the behavior of target actors (e.g., violation of obligations, claims, or promises) up to public scrutiny. An article that appeared in *The Washington Post* on May 31, 1998 (and was posted to the list serve on the same date), entitled "Canada Failing to Enforce Environmental Laws, Reports Find," alleges: "Canada is neglecting to enforce the environmental commitments it led the world in creating ... the government is not keeping the promises it makes both to Canadians and to the world."[26] (Schneider

1998). By publicly exposing and shaming the Canadian government for its alleged failure to keep its environmental commitments, the network attempts to provoke and direct approbation into a significant source of pressure that can be applied to influence subsequent governmental action.

Networks use material leverage when they link issues to capital or to votes, political office, or other benefits. To enable effective use of this technique, targets must be vulnerable to material incentives, to sanctions or to various forms of pressure, such as public opinion (Keck and Sikkink 1998). Leverage is exerted through the ballot box when networks affect public opinion through various media or, in the case of large networks, through mobilization of their own membership.

Y2Y often posts lobbying opportunities on its list serve to allow members to utilize this form of influence. For example, a June 1998 "action alert" provided network members with details on the approval of a proposal to construct a seven-story, 156,000-square-foot conference center near the shores of Lake Louise in Banff National Park. Members were encouraged to contact the Minister of Canadian Heritage, the Secretary of State (Parks), and the Prime Minister to take a stand on the issue of development in Canada's national parks (unpublished communication, Y2Y list serve, June 16, 1998). Another call for lobbying was issued on July 13, 1998, reminding network members to "call the Farm Bureau today to express your desire that the Farm Bureau drop its lawsuit to remove wolves from Yellowstone and Idaho and to stop their harassment of wild wolves" (unpublished communication, Y2Y list serve, July 14, 1998). Cyberspace promotion of lobbying opportunities is becoming more commonplace. On May 19, 1998, list serve members were advised that a political action site had been added to the Internet[27] to "assist Canadians in communicating with their elected representatives."

The levers of power can also be located in the economic realm. Groups can focus on corporations and pressure them through the release of damaging information about environmentally destructive practices, thereby mobilizing public outrage. For example, a news release distributed over the Y2Y list serve on June 30, 1998, disclosed that the Banff Environmental Action and Research (BEAR) Society, a Y2Y network member, had just launched an international postcard campaign to curtail the death toll of large mammals on the Canadian Pacific Railroad (CPR) rail line through Banff and Yoho National Parks. From 1991 to 1998, 523 large animals were killed on the main line of the CPR, a giant multinational corporation with record profits of 1.25 billion in 1997. Juxtaposed on the postcard with the photograph of a black bear that had been decapitated on the CPR main line in Yoho National Park is: "Record Profits, Record Losses." The reverse of the postcard offers a letter to Canadian Prime Minister Jean Chrétien, asking him to "keep his promise to 'protect Banff National Park for citizens of Canada and the world—forever'" (unpublished communication, Y2Y list serve, June 30, 1998).

Accountability Politics Accountability politics are pursued when activists use their command of information and the discursive positions taken by policymakers to expose the distance between discourse and practice (Keck and Sikkink 1998). The Friends of the Oldman River, a Y2Y member group, challenged the Canadian Department of Fisheries and Oceans' use of directives, rather than legislation or regulations, to protect water resources and fisheries. The group posted the following update to the list serve on July 21, 1998:

The Government of Canada argues that it "is effectively enforcing its environmental laws and is ... in full compliance with its obligations under the NAAEC [North American Agreement on Environmental Cooperation]." However, Canadian Federal Court Justice Muldoon commented: [T]his is a transparent bureaucratic attempt at sheer evasion of binding statutory imperatives. By making 'policy' not contemplated by the statutes, the DFO types simply cannot immunize the Minister and DFO from judicial review, nor circumvent the environmental laws which they decline to obey.

Information may also be used to hold policymakers or implementing agencies responsible for perceived violations of the public trust. An article from the *Idaho Falls Post Register* (Trillhaase 1998) was posted to the list serve on June 22, 1998. The article alleges that when a prominent Idaho rancher's grazing privileges on public land were revoked, he sought compensation for lost "rights" through the Bureau of Land Management. When this attempt failed, the rancher, who has "given generously to Republican campaigns" in the past, "turned to his friends in the Senate," who drafted a bill to compensate the rancher, in the amount of

$264,750, for lost grazing rights—grazing rights that generated only $1,429 annually for the government. Moreover, the article continues, the true amount of the compensation could reach $1 million "after the government pays the cost of replacing water pipes and fencing on whatever new grazing lands [the rancher] turns to." "The trouble is," the article concludes, the senators "aren't playing with their own money and their own land. They're playing with yours."

Conclusions: Information and Communication Technologies and Discourse Diffusion

In summary, recent advances in scientific and communication technologies both enable the growth of networks like Y2Y and shape the strategic repertoires through which they function. Facilitated by advances in scientific knowledge and computer technology,[28] Y2Y uses its command of information, imagery, and symbol to build and link a coalition of geographically distant groups, to aid in the development of discursive frames, strategies, and policy proposals, and to mobilize and coordinate action among its adherents. Through the provision of various kinds of information, problem definitions, or strategic framing devices, as well as models for cooperative action, training opportunities, and networking tools, Y2Y facilitates the efforts of grassroots actors. Grassroots adherents use these tools to advance their local agendas, to initiate or nurture value change in the local communities, to generate constituencies for policy proposals, to form coalitions with other groups, and to disseminate the message of Y2Y.

The command of information also allows network members to target political and corporate decision makers directly or indirectly and to provide them with various types of evidence in favor of their discursive constructions of environmental problems and solutions. The potential to mobilize the outrage and/or action of various segments of the public allows the network to increase decision makers' awareness of the costs and benefits associated with alternative policy proposals or actions. This may lead to a change in perceived interests, attitudes, or values among powerful government or economic actors. As a final note, the ability to expose the positions and actions of political and economic actors also increases their accountability to the community at large.

Similar to the Lake Constance networks studied by Blatter, Y2Y represents a "best-case scenario" for successful transnational environmental networking in many ways. It boasts a community of dedicated and determined activists with extensive experience in networking, familiarity with social and political processes, and a long history of involvement in environmental campaigns. Network actors share similar worldviews, problem definitions, action paradigms, and legitimating national discourses, as well as a common language and a long history of relatively amicable cooperation among nations. Although cultural and ideological differences do exist, unlike in the Georgia/Abkhazia case study presented by Garb and Whiteley, they are not acute or intractable enough to represent barriers to successful cooperation.

In comparison with environmental activists in other regions, the network is relatively affluent in terms of technological, scientific, and financial resources. Y2Y actively determines gaps in social, political, or scientific knowledge, finds ways to fill the gaps, and incorporates the information into its working strategy.[29] Y2Y ties its claims to scientifically legitimated goals of biodiversity preservation that have been recognized as an international goal by institutions such as the Brundtland Commission and the Rio Declaration and articulated in national discourses or statutes in both Canada and the United States. The initiative has also received the acclamation of the international community. Y2Y was recognized as one of the "leading edge conservation efforts on the planet" at a 1997 World Meeting of the International Union for the Conservation of Nature (IUCN), a factor that may weigh heavily in determining the success of the initiative.[30] Obviously, international networking has resulted in significant gains for the Y2Y network.

International networking also presents difficulties, however. Kriesberg (1997) states that operating internationally increases transaction costs. Direct participation in international networks is limited to individuals or group members with sufficient financial resources and the freedom to travel. The Y2Y network offers limited financial assistance when possible and changes venues to enable fuller participation; however, participation for the majority typically occurs only within the realm of cyberspace. Most active Y2Y members are dedicated activists with very different personalities, skills, working styles, strategy preferences, and opinions on how best to pursue particular common goals. Finding strategies all members can endorse will be critical to achieving a cohesive approach. For these and other reasons previously examined, the development of an efficient, accessible, and effective communication system assumes paramount importance in international networking.

We have seen that the involvement of a population of diverse grassroots activists vastly facilitates information gathering and the diffusion of ideas through the environmental community and within the larger community as well. The group's diversity translates into flexibility, resourcefulness, and adaptability, all of which are critically important when a wide range of strategic approaches is mandated because of uncertainty about the chances for success in any single venue. Diversity also means that linkages with other key players or groups may be more easily established. In addition, the wide range of experience and skills held by group members increases the probability that more ideas will be fielded, a wider range of initiatives and options will be explored and pursued, and more problems will find solutions. In sum, it may indeed be true that, in the words of Harvey Locke (Yellowstone to Yukon Conservation Initiative 1997, 16), "If we cannot save wildlife and wilderness in the Yellowstone to Yukon region of Canada and the United States, there is little hope for either in the world."

As a future goal, Y2Y proposes the creation of roundtable forums in which critical social, economic, scientific, and cultural information would be elicited and exchanged, leading to better-informed land, water, and resource use decisions. Ongoing social interaction and cooperation within the roundtables might offer a mechanism for integrating the diverse needs and perspectives of the panoply of social actors who are now contending environmental issues and foster development of a consensual body of knowledge in favor of preservation and long-term sustainability and security.

However, ecological values do not always fare better in a public arena constructed around the incorporation of diverse interests through ongoing discourse, as a recent example in the Alberta sector of Y2Y illustrates. A citizens' "environmental panel," comprising environmental, industrial, recreational, agricultural, and indigenous interests, recommended in May 1998 that logging, mining, and oil drilling should be permitted in the ecologically sensitive and critically important habitat represented by the Whaleback region of the Rocky Mountains (Mitchell 1998).

Therefore, the establishment of this type of forum on its own is insufficient to achieve Y2Y's goals of sustainability, watershed protection, landscape connectivity, and biodiversity preservation without concomitant, commendatory value change in the general public. Historic and contemporary patterns of extraction, exploitation, production, and consumption continue to threaten natural systems—and the human communities within them—around the world. Decisions such as that of the Whaleback citizens' committee are often the result of external economic pressures that encourage social actors to sacrifice long-term environmental goals to short-term economic incentives (Lipschutz 1996). However, if land and water become integrated with globally recognized values such as the preservation of environmental and human security, their community value may come to exceed their commodity value, resulting in preferences for preservation rather than exploitation. Re-integrating land and water with these universalistic values, therefore, offers a way of increasing community resistance to external economic pressures, or to what Starke (1990, 28) calls the "tyranny of the immediate."

Keck and Sikkink (1998) remind us that transnational advocacy networks often do not achieve their goals, adding that environmental campaigns that have had the greatest transnational effects have stressed the connection between protecting environments and protecting the vulnerable people living in them. Y2Y may or may not achieve full success in terms of the group's measures; it may instead be successful in terms of creating a transnational public sphere in which issues of economic and social justice may be debated and in which marginalized actors may be empowered. Putnam, Leonardi, and Nanetti (1993) refer to this outcome as the creation of social capital: a combination of consciousness, networks, and trust that permits the community to act collectively toward the common good. Y2Y's challenge, then, is to foster an understanding of environmental preservation as the common good. Y2Y uses its command of information, image, and symbol and its extensive network of grassroots and other alliances to provide communities with the kind of knowledge they need to resist the external forces of exploitation and make the transition toward a more sustainable and equitable future. In so doing, Y2Y may reinvigorate local communities that are, in turn, the building blocks of the Y2Y movement.

Historic land, water, and resource use paradigms in the Y2Y region have been thrown into question,[31] and evolving paradigms increasingly emphasize noncommodity values such as biodiversity preservation or heritage, cultural, spiritual, and recreational benefits. Scientific information from the biologic, geologic, social, and environmental sciences is actively restructuring cognitive maps and discourses related to the nature-human relationship. The integration of these new constructions of environmental phenomena and relationships into everyday worldviews and practices ultimately may allow us to address the problems of the environment (Lipshutz 1996).

The Y2Y initiative is itself a tapestry—a tapestry woven from the legacy of the past, the encroachments of modernity, and the intention to create a sustainable future. The network re-imagines the landscape and the community in new and innovative ways that are firmly grounded in cutting-edge economic, social, and scientific research. Y2Y finds its voice in the creation and transboundary diffusion of a nature-centered discourse that articulates a new way for the human community to relate to land, to water, and to the diversity of life they nurture.

Notes

1. I would like to acknowledge and thank Matt Reid, George Smith, Craig Stewart, and Kathleen Wiebe for their assistance in providing information or for their editorial comments on previous renditions of this manuscript.

2. Conservation biology addresses the biology of species, communities, and ecosystems disturbed by human activities or other agents, either directly or indirectly. Its goal is to provide the principles and tools for preserving biological diversity. Conservation biology is considered to be a "crisis discipline," in which decisions must often be made before complete scientific information is available (Noss 1994, 35).

3. Monte Hummel and Arlin Hackman (1994, xviii) of the World Wildlife Fund, Canada, have defined biodiversity as "the variety of life [that] includes the full

range of genes, species, communities, ecosystems, functions, and evolutionary processes."

4. For example, it is now feared that Yellowstone's geysers are threatened by the development of geothermal energy outside the park's borders, and the park's flagship animal—the grizzly bear—is believed imperiled by development and other human-caused threats beyond Yellowstone's borders (Primm and Clark 1996, 137).

5. In addition to editing the *Journal of Conservation Biology*, Noss acts as science editor for *Wild Earth*, the journal of The Wildlands Project, and is an international consultant in conservation and an adjunct professor at Oregon State University.

6. Conservation biology, landscape ecology, and island biogeography are considered to be holistic, multidisciplinary sciences; that is, they derive their premises, techniques, and tools from multiple disciplines. (For more information on conservation biology, see Noss 1994.)

7. Gary Tabor of Center for Conservation Medicine in Andover, Massachusetts, is the author of *Yellowstone-to-Yukon: Canadian Conservation Efforts and a Continental Landscape/Biodiversity Strategy*, a report commissioned on behalf of Y2Y by the Henry P. Kendall Foundation. Tabor trained as a wildlife veterinarian at the University of Pennsylvania and in conservation biology at the Yale University School of Forestry.

8. Dave Foreman is the publisher of *Wild Earth* and the chair and cofounder of The Wildlands Project, on which the Y2Y concept is based. He has served on the board of the Sierra Club and has worked as a lobbyist for the Wilderness Society. Foreman is also the founder of Earth First!

9. For the sake of brevity, I use the term "land" throughout this document in the sense ascribed to it by Aldo Leopold (1949). In Leopold's conceptualization, which has been adopted by Y2Y, land includes "soils, waters, plants, and animals."

10. David Mayhood and Michael Sawyer assert that although all ecosystems are subjectively defined, some definitions make more sense ecologically. Functionally, according to Mayhood and Sawyer, "ecosystems are dynamic interactions of their abiotic and biotic components, which display considerable spatial and temporal variation." The Y2Y bioregion was delineated on the basis of conceptual and practical manageability: too large an area would lead to difficulties in achieving consensus on conservation issues, too small an area to fit the needs of the wide-ranging species under consideration would serve little purpose in achieving the network's goals. In reality, it has been agreed that Y2Y has "no discreet boundaries except those designed by people.... The people ... who have coalesced around the Y2Y idea did so for subjective reasons, not because Y2Y is a scientific unit unrelated to their personal worldview. If we fail to acknowledge this, we will be transparently dishonest" (Mayhood and Sawyer 1997, 5–6).

11. Wilcox is the coordinator of the Sierra Club Grizzly Bear Ecosystems Project and a board member of The Wildlands Project. She has also served as project

director for an information-gathering effort on the physical characteristics, species distribution, natural processes, human influences and trends and approaches to conservation in the Y2Y region. For the findings of this effort, see Harvey 1998.

12. Butler, a professor of Parks, Forestry and Conservation Biology at the University of Alberta, has been a leading voice in the international conservation movement for more than twenty years. He is also a Buddhist monk.

13. Participants in the meeting included David Johns, Michael Soulé, and John Davis of The Wildlands Project; Tim Clark; Reed Noss; Monte Hummel and Arlin Hackman of World Wildlife Fund; John Weaver and Louisa Wilcox, scientists and activists from Montana; and Steve Herrero from the University of Calgary. The meeting was held in Kananaskis, Alberta, Canada, in December 1993 and was hosted by CPAWS.

14. One of Y2Y's first tasks was to conduct a "circuit ride" throughout the U.S. portion of the proposed study area. Presentations on the Y2Y concept were delivered to sixty-three conservation groups in thirty-one communities in Idaho, Washington, Wyoming, and Montana. Support for Y2Y was gauged, and there was found to be near universal approval among the conservation groups for the idea. The same type of circuit ride is now being planned for the Canadian portion of Y2Y (Matt Reid, personal communication, October 11, 1998).

15. At this writing, the coordinator, assistant to the coordinator, and outreach coordinator are the only paid positions in the network; all other positions are filled by volunteers.

16. For information on the hike, visit the hikers' Internet site at ⟨http://www.y2y@rockies.ca⟩.

17. For a discussion of Banff National Park, see Davis 1995. See also Chase 1987 for a similar discussion of Yellowstone.

18. Jim Morrison is a Canadian conservationist, author, and ethnohistorian with extensive experience in aboriginal land claims and policy. He has served as a consultant to numerous First Nations tribes over the course of the last twenty years (Hummel 1995).

19. The CCEDA itself represents a transnational effort, instigated by Denis Gourdeau of CPAWS, to facilitate the adoption of ecosystem-based management practices in the Y2Y region. Although researchers in the United States (principally located at the University of Montana) supported the effort at the outset, the initiative eventually reverted to being primarily a Canadian undertaking (Kevin Van Tighem, Conservation Biologist at Waterton Lakes, personal communication, November 24, 1997).

20. Bart Robinson, former editor of *Equinox* magazine (a Canadian publication), has an extensive background in print journalism and publications, writing, editing, and administration.

21. Contributors from within the network include the Alberta Wilderness Association, British Columbia Wild, the Canadian Nature Federation, CPAWS

(Calgary/Banff and Yukon chapters), Friends of the Northern Rockies, Predator Project, Save-the-Cedar League, The Wilderness Society (Northern Rockies region), The Wildlands Project, and an anonymous donor.

22. Main headings in the report include "Yellowstone to Yukon: A Physical Overview," "Human Influences and Trends," Conservation, Species and Natural Processes," "Approaches to Conservation," and "A Summary of Issues Facing the Yellowstone to Yukon."

23. The report, funded by the Henry P. Kendall Foundation and completed in July 1996, details the state of landscape protection along the Central and Southern Canadian Rocky Mountains region of Y2Y and describes the problems and threats facing wilderness and biodiversity preservation in the region. It also discusses past and current protected area strategies that have been undertaken at the federal and provincial levels and the impacts of previous policies, among other topics.

24. The theme of inequitable standards for environmental preservation between the developed and developing worlds is carried through in another newspaper piece that points out that Y2Y does not propose to stop development or evict humans. Instead, the article continues, Y2Y argues that "two of the world's richest economies should be able to conserve a portion of its [North America's] remaining wilderness" (Schneider 1997).

25. Of the approximate eleven-million-acre designation, extractive interests strictly protect only 2.5 million acres from any exploration or development. The remaining acreage is set aside as "special management zones" in which exploration and development may be allowed within certain relatively strict limitations based on environmental sustainability and the preservation of wilderness and wildlife habitat. In addition, all roads must be removed at closure of activities.

While the network has been greatly heartened by the protection of the Muskwa-Kechika lands, this victory was achieved in a remote and relatively uninhabited region under the auspices of a reasonably environmentally progressive government and was achieved only at the cost of concessions in which environmentalists agree not to contest resource development in other (specified) areas, elsewhere. It is unlikely that victories will come so easily or be so extensive in other regions of Y2Y.

26. In another example, network members were advised that a recent report on forest practices in British Columbia was available on the Sierra Club Legal Defense Fund's Web site. The Forestry Report Card, prepared by several of the province's prominent environmental groups and coordinated with the World Wildlife Fund's Endangered Spaces Progress Report, assigns British Columbia a grade of D in four forest practices criteria and an F in the remaining eight. "This report card exposes the shell game the timber industry has been playing," Dave Neads, chairperson of the Forest Caucus of the British Columbia Environmental Network stated, adding, "They have been telling the world that British Columbia has world-class practices, while, at the same time, forcing government to gut

these standards, systematically removing the flimsy environmental safeguards that once existed" (unpublished coarse-filter communication, April 30, 1998).

27. The site, managed by the Political Action Committee at Earth-Nexus, is located at ⟨http://www.earth-net.net/pac/canada/federal/index.html⟩.

28. Mapping of watersheds, geophysical characteristics, and animal movement patterns has been greatly facilitated by satellite (Landsat) imagery, powerful computer equipment and software, and Geographic Information Systems (GIS) technologies. These technological advances facilitate the increased emphasis on eco-regions/bioregions and watersheds by public land management and parks personnel. Computer models simulate animal movement patterns based on data that can be aggregated and easily accessed in centralized databases. Radio-collaring technology has greatly increased scientific knowledge about habitat needs and animal behaviors. The ability to map animal movements and home ranges has led to an increased awareness of the importance of contiguous wild-land matrices and guided the development of the model on which Y2Y is based. The scientific finding that certain large animals, such as wolves and bears, ranged all the way from the Yukon to Yellowstone Park is a key factor in the formation and strategic approach of Y2Y.

29. For example, Y2Y is working closely with The Sonoran Institute to develop workable regional plans for Y2Y. The institute's community-based approach to conservation incorporates economic and social needs into designs that local people then develop, support, and implement (Sonoran Institute 1996).

30. Information presented during the public showing of "Reweaving the Wild" at the Southern Alberta Jubilee Auditorium, May 4, 1998. Dave Foreman and Harvey Locke were keynote speakers.

31. For a further discussion of changing paradigms in the Y2Y region, see Grumbine 1994 and Environment Canada 1986.

References

Barker, Mary L., and Dietrich Soyez. 1994. "Think Locally, Act Globally? The Transnationalization of Canadian Resource Use Conflicts." *Environment* 36, no. 5 (June): 12–24.

Blowers, Andrew, and Pieter Leroy. 1995. "Environment and Society: Shaping the Future." In Pieter Glasbergen and Andrew Blowers (eds.), *Environmental Policy in an International Context: Perspectives on Environmental Problems.* London, Sydney, Auckland: Arnold, pp. 255–283.

Brysk, Alison. 1994. "Acting Globally: Indian Rights and International Politics in Latin America." In Donna Lee Van Cott (ed.), *Indigenous Peoples and Democracy in Latin America.* New York: St Martin's Press, pp. 29–51.

Brysk, Alison. 2000. *From Tribal Village to Global Village: Indian Rights and International Relations in Latin America.* Stanford, Calif.: Stanford University Press.

Butler, Jim. 1997. "How to Effectively Defend the Earth While Maintaining Your Optimism, Energy, Sanity, Sense of Humour, Family and Keep Your Dog from Growling at You." In Stephen Legault et al., eds., *Connections: Proceedings from the First Yellowstone to Yukon Conservation Initiative*, Waterton-Glacier International Peace Park, October 2–5. Canmore, Alberta, Canada: Yellowstone to Yukon Conservation Initiative, pp. 65–78.

Castells, Manuel. 1996. "Conclusion: Social Change in the Network Society." In *The Rise of the Network Society*. Cambridge, Mass.: Blackwell Publishers, pp. 354–362.

Chase, Alston. 1987. *Playing God in Yellowstone: The Destruction of America's First National Park*. Orlando, Fl.: Harcourt Brace Jovanovich.

Clark, Tim W. 1993. "Creating and Using Knowledge for Species and Ecosystem Conservation: Science, Organizations, and Policy." *Perspectives in Biology and Medicine* 36, no. 3 (Spring): 497–525.

Clark, Tim W. 1996. "Learning as a Strategy for Improving Endangered Species Conservation." *Endangered Species Update* 13, nos. 1 & 2: 5 (5 pages, nonsequential).

Clark, Tim W. 1997. *Averting Extinction: Reconstructing Endangered Species Recovery*. New Haven, Conn., and London: Yale University Press.

Clark, Tim W., and Richard L. Wallace. 1998. "Understanding the Human Factor in Endangered Species Recovery: An Introduction to Human Social Process." *Endangered Species Update* 18, no. 1.

Clark, Tim W., et al. 1992. "Conserving Biodiversity in the Real World: Professional Practice Using a Policy Orientation." *Endangered Species Update* 9, nos. 5 & 6.

Clark, Tim W., and David L. Gaillard. n.d. "Organizing an Effective Partnership for the Yellowstone to Yukon Conservation Initiative."

Davis, Glen. 1995. "Playing God in Endangered Spaces: Perspectives of a Donor." In Monte Hummel (ed.), *Protecting Canada's Endangered Spaces: An Owner's Manual*. Toronto: Key Porter Books, pp. 181–189.

Environment Canada. 1986. *Survival in a Threatened World*. Submission by the People of Canada to the World Commission on Environment and Development, May, 1986. Victoria: Environment Canada.

Foreman, Dave. 1997. "Yellowstone to Yukon and the Wildlands Project." In Stephen Legault et al. (eds.), *Connections: Proceedings from the First Yellowstone to Yukon Conservation Initiative*, Waterton-Glacier International Peace Park, October 2–5. Canmore, Alberta, Canada: Yellowstone to Yukon Conservation Initiative, pp. 9–21.

Grumbine, R. Edward, ed. 1994. "Introduction." In *Environmental Policy and Biodiversity*. Washington, D.C.: Island Press, pp. 3–19.

Haas, Peter M. 1992. "Epistemic Communities and International Policy Coordination." *International Organization* 46, no. 1 (Winter): 1.

Harvey, Ann, ed. 1998. *A Sense of Place: Issues, Attitudes and Resources in the Yellowstone to Yukon Region*. Canmore, Alberta: Yellowstone to Yukon Conservation Initiative.

Hughes, David. 1998. "Mapping the Mozambican Environment: Territorial Politics on the Frontiers of Enclosure." Paper presented to the conference "The Environment in Context: Democracy, Capitalism and Culture," University of California at Irvine, June.

Hummel, Monte. 1995. "Author Biographies." In Monte Hummel, ed. *Protecting Canada's Endangered Spaces. An Owner's Manual*. Toronto: Key Porter Books, pp. 231–235.

Hummel, Monte, and Arlin Hackman. 1994. "Introduction." In Monte Hummel, ed., *Protecting Canada's Endangered Spaces: An Owner's Manual*. Toronto: Key Porter Books, pp. xi–xix.

Johns, David. 1997. "Conference Synthesis." In Stephen Legault et al., eds., *Connections: Proceedings from the first Yellowstone to Yukon Conservation Initiative*, Waterton-Glacier International Peace Park, October 2–5. Canmore, Alberta, Canada: Yellowstone to Yukan Conservation Initiative, pp. 60–64.

Keck, Margaret, and Kathryn Sikkink. 1998. *Activists beyond Borders*. Ithaca, N.Y.: Cornell University Press.

Kriesberg, Louis. 1997. "Social Movements and Global Transformation." In Jackie Smith et al. (eds.), *Transnational Social Movements and Global Politics: Solidarity Beyond the State*, pp. 3–18. Syracuse, N.Y.: Syracuse University Press.

Leopold, Aldo. 1953. *A Sand County Almanac*. New York: Ballantine Books.

Legault, Stephen, et al. 1997. "Building Regional Networks. Putting Yellowstone to Yukon on the Ground." In Stephen Legault et al. (eds.), *Connections: Proceedings from the First Yellowstone to Yukon Conservation Initiative*, Waterton-Glacier International Peace Park, October 2–5. Canmore, Alberta, Canada: Yellowstone to Yukon Conservation Initiative, pp. 79–95.

Lipschutz, Ronnie D. 1996. *Global Civil Society and Global Environmental Governance: The Politics of Nature from Place to Planet*. Albany: State University of New York Press.

Locke, Harvey. 1996. "Saving Wildlife from Yellowstone to Yukon." *Wildlife Conservation* 99, no. 5 (September–October): 24.

Locke, Harvey. 1997. "The Yellowstone to Yukon Vision." In Stephen Legault et al. (eds.), *Connections. Proceedings from the First Yellowstone to Yukon Conservation Initiative*, Waterton-Glacier International Peace Park, October 2–5. Canmore, Alberta, Canada: Yellowstone to Yukon Conservation Initiative, pp. 33–48.

Mayhood, David, and Michael Sawyer. 1997. "Fishes of the Yellowstone to Yukon." In *A Sense of Place: Issues, Attitudes and Resources in the Yellowstone to Yukon Region*, Summary of the Interim Draft, prepared for the Y2Y Connections Conference, October 2–5. Canmore, Alberta, Canada: Yellowstone to Yukon Conservation Initiative, pp. 5–7.

McAdam, Doug, et al. 1996. "Introduction: Opportunities, Mobilizing Structures, and Framing Processes—Toward a Synthetic, Comparative Perspective on Social Movements." In Doug McAdam, John D. McCarthy, and Mayer Zald (eds.), *Comparative Perspectives on Social Movements: Political Opportunities, Mobilizing Structures, and Cultural Framings.* New York: Cambridge University Press, pp. 1–22.

McCarthy, John D. 1997. "The Globalization of Social Movement Theory." In Jackie Smith et al. (eds.), *Social Movements and Global Politics: Solidarity beyond the State.* Syracuse, N.Y.: Syracuse University Press, pp. 243–259.

Mitchell, Alanna. 1998. "Rare Rockies Wilderness at Risk: Environmental Panel Says Alberta's Whaleback Should Be Drilled, Mined, Logged." *Globe and Mail,* Alberta Bureau, May 20.

Morrison, Jim. 1995. "Aboriginal Interests." In Monte Hummel (ed.), *Protecting Canada's Endangered Spaces: An Owner's Manual,* pp. 18–26. Toronto: Key Porter Books.

Noss, Reed. 1994. "The Wildlands Project: Land Conservation Strategy." In Edward R. Grumbine (ed.), *Environmental Policy and Biodiversity.* Washington, D.C.: Island Press, pp. 233–266.

Noss, Reed. 1992. "The Wildlands Project: Land Conservation Strategy." *Wild Earth Special Issue.* Canton, New York: The Cenozoic Society, pp. 10–24.

Peepre, Juri. 1995. "Yukon." In Monte Hummel (ed.), *Protecting Canada's Endangered Spaces: An Owner's Manual,* pp. 153–162. Toronto: Key Porter Books.

Potter, David. 1995. "Non-governmental Organizations and Environmental Policies." In Pieter Glasbergen and Andrew Blowers (eds.), *Perspectives on Environmental Problems.* London, Sydney, Auckland: Arnold, pp. 25–49.

Primm, Steven A., and Tim W. Clark. 1996. "The Greater Yellowstone Policy Debate: What Is the Policy Problem?" *Policy Sciences* 29, no. 2 (May): 137–166.

Putnam, Robert D., Robert Leonardi, and Rafaella Y. Nanetti. 1993. *Making Democracy Work: Civic Traditions in Modern Italy.* Princeton, N.J.: Princeton University Press.

Rasker, Ray, and B. Alexander. 1997. *The New Challenge: People, Commerce and the Environment in the Yellowstone to Yukon Region.* Report commissioned on behalf of the Yellowstone to Yukon Conservation Initiative by The Wilderness Society. Washington, D.C.: The Wilderness Society.

Schindler, David W. 1997. "Aquatic Issues in the Yellowstone to Yukon." In *A Sense of Place: Issues, Attitudes and Resources in the Yellowstone to Yukon Bioregion, Summary of the Interim Draft,* prepared for the Y2Y Connections Conference, Waterton Lakes, Alberta, Canada, October 2–5.

Schneider, Howard. 1997. "Conservationists Take Stock of the Land. Group Envisions Protected Area from Yellowstone to Yukon." *Washington Post,* October 27, A20.

Schneider, Howard. 1998. "Canada Failing to Enforce Environmental Laws, Reports Find." *Washington Post*, May 31, A26.

Slater, David. 1997. "Terrains of Power, Movements of Resistance." Paper presented at the 1997 meeting of the Latin American Studies Association, Continental Plaza Hotel, Guadalajara, Mexico, April 17–19.

Smith, Jackie. 1997. "Characteristics of the Modern Transnational Social Movement Sector." In Jackie Smith et al. (eds.), *Transnational Social Movements and Global Politics: Solidarity beyond the State*. Syracuse, N.Y.: Syracuse University Press, pp. 42–58.

Smith, Jackie, et al., eds. 1997. *Transnational Social Movements and Global Politics: Solidarity beyond the State*. Syracuse, N.Y.: Syracuse University Press.

Sonoran Institute. 1996. "Partners in Community Stewardship." 1996 Annual Report. Tucson, Ariz.: Sonoran Institute.

Starke, Linda. 1990. *Signs of Hope: Working toward Our Common Future*. Oxford: Oxford University Press.

Tabor, Gary. 1996. *Yellowstone-to-Yukon: Canadian Conservation Efforts and a Continental Landscape/Biodiversity Strategy*. Report commissioned on behalf of Y2Y and printed by the Henry P. Kendall Foundation, Boston, Massachusetts.

Trillhaase, Marty. 1998. "The Best Money Can Buy." *Idaho Falls Post Register*, June 14, 1998. Available: ⟨http://www.idahonews.com/061498/OPINION/20637.htm⟩.

Wilcox, Louisa. 1998. Statement made during the Y2Y Council meeting, Helena, Montana, April 24–26, 1998.

Wilson, Jeremy. 1992. "Green Lobbies: Pressure Groups and Environmental Policy." In Robert Boardman (ed.), *Canadian Environmental Policy: Ecosystems, Politics, and Process*. Toronto, Oxford, New York: Oxford University Press.

Yellowstone to Yukon Conservation Initiative. 1997. *A Y2Y Network Handbook: Principle, Policies and Guidelines*. Adopted April 27, 1997. Canmore, Alberta: Author.

Yellowstone to Yukon Conservation Initiative (nd). The Yellowstone to Yukon Conservation Initiative. To Restore and Protect the Wild Heart of North America. (Brochure.) Canmore, Alberta, Canada: Yellowstone to Yukon Conservation Initiative.

6

Discursive Practices and Competing Discourses in the Governance of Wild North American Pacific Salmon Resources

Kathleen M. Sullivan

The Salmon Wars. This evocative headline has recurred frequently in United States and Canadian newspapers and magazines in recent years. As each new commercial salmon fishing season approaches, vigorous public debates about how to best manage salmon resources and who should capture how many salmon erupt in the Pacific Northwest.

This chapter[1] explores one prominent transnational debate over water resources: the mass-mediated, international struggle over wild North American Pacific salmon, with a special emphasis on the events during the summer of 1997. Rather than being uniquely spectacular, these events reflect and symbolize contemporary debates over salmon specifically and transnational water resources in general. The goal of this chapter is to unpack the discursive practices in the "salmon wars" in a theoretically specific way. I aim to focus attention on the underpinning systems of knowledge created and deployed in this transnational struggle over water resources. My analysis is motivated by the observation that these systems of knowledge shape transnational environmental governance, economic production, and the material environmental conditions in which people must live and work. How might the plurality of voices engaged in producing social knowledge about salmon and water resources, which in turn affects people's lives, be fostered? This chapter suggests that the necessary first step is to analyze the various discursive representations available and consider the ramifications and limitations of these discursive practices. Chapter 7 also considers how we might use discourse analysis to enhance the plurality of voices engaged in a transborder water resource issue.

The regional struggle over wild salmon transcends national boundaries and illustrates the tensions between the global and the local, "glocalization," described in chapter 1. The mix of water and salmon engenders complex economic and environmental relationships that challenge the supremacy of the territorially defined nation-state. Salmon require extensive fresh-marine water systems for their survival. Consequently North American Pacific salmon migrate through a number of different fisheries production regimes as well as fisheries management regimes during their lives. As human pressures on the marine environment have increased through time, fisheries management regimes have become ever more comprehensive and stringent. Timbering, dams, more efficient commercial and recreational fishing practices, urbanization, and resource management practices challenge the integrity of the interdependent fresh-marine water systems and the livelihoods of everyone dependent on those fresh-marine water systems species. These new constellations of exploitation are reconfiguring the relationships between the local, the national, and the transnational in very complex ways. The nation-state faces challenges to its sovereignty, but at the same time, many of the challenging discursive practices actually reconfirm that sovereignty.

The environmental and economic tensions between the United States and Canada over catches of wild salmon have garnered regional and national press coverage. Currently, the Pacific Salmon Treaty[2] is the pivotal instrument for resolving U.S.-Canadian conflicts over wild salmon. Repeated failures over the last several decades to negotiate and abide by the treaty have become a focal point in vociferous public debates about wild salmon. Social actors contribute to the public debates by anchoring their arguments in discursive themes of conservation, economic equity, and nationalism.

Fisheries also provide an informative perspective from which to examine the issues about water raised in chapter 2. Fish and their watery habitats ignore international boundaries. Like flowing water, anadromous fish are a fugitive resource, difficult to secure and control. Fisheries, which are human social systems, depend on fish and their watery habitats. Fisheries illustrate the way water and its social definitions and uses are bound to the survival of all life forms. Certainly from a fisheries management perspective, water is a "gift of nature." The vulnerability of both fish stocks to overfishing and fish habitats to destruction suggest that there are ecological limits to human exploitation. The economic value of fisheries and the national ties of fishing fleets suggest the persistence of modern perspectives of commodity and property. But from another perspective, that of fishers, fishing is not just a way of making a living; it is a way of life imbued with dignity. Their perspective suggests that there is a bond of water and culture that is often underemphasized in the modernist commodity perspective.

My examination of the salmon wars draws on the work of Michel Foucault and Jürgen Habermas. Although their approaches are very different, each is concerned with the construction of knowledge in modern society. My overarching theoretical strand engages Foucault's notion of a discursive field to unpack the discursive themes and relations of power in the public debates over wild salmon resources. Foucault's work suggests that all knowledge is social and created through the deployment of discursive elements, including discursive themes, in the modern encompassing network of relations of power.

I then turn to Jürgen Habermas's richly suggestive notion of the colonization of the public sphere (or forum) through the use of instrumental, technocratic rationality. This act of colonization refers to the pervasive practice of reducing complex social problems to technical ones and thereby eliminating the plurality of knowledge about an issue. Generally Habermas devotes very little attention to the constitutive nature of exercises of power in the contemporary social landscape. However, his critique of instrumental action exposes exercises of power that inhibit the plurality of voices in the public forum.

The following section examines the key discursive themes of "conservation," "economic equity," and "nationalism" that resonate, reinforce, and contradict each other as they organize the public debates over salmon and their watery habitats. The second section examines the discursive strategies of various social actors as the mass media represent them. The conclusion reconsiders the contradictory implications of globalizing discourses and systems of knowledge that remain embedded in modernist notions of nation-state sovereignty and a capitalist economic formation.

Discursive Fields, Discourses, and Themes: Conservation, Economic Equity, and Nationalism

The production of knowledge in contemporary society includes what is said, what can be said, and what is silenced. Discourse analysis describes and analyzes the discursive practices of different social actors as they interact with each other. We can excavate the systems of knowledge that shape environmental issues, regulatory policies, negotiations over responsibility, economic development, and economic rights to water resources. Theoretically driven inquiry into the processes of creating knowledge and the authority to wield knowledge exposes the assumptions and often-unnoticed social practices underpinning the salmon wars. We can then reflect on those practices and their ramifications.

To excavate these influential and deeply rooted assumptions, I turn to Michel Foucault (1972). He directs our attention to the notion of discursive fields rather than simply to discourses. The theoretical notion of discourse has been used with increasing frequency in the sociological and political scientific study of environmental and ecological issues.[3] However, not all of these analyses define and use the notion of discourse in the same way, and some do not allow us to consider fully the role of relations of power among various social actors. If our goal is to enhance the plurality of voices in the public forum, then consideration of exercises of control and power is of paramount importance. The notion of the discursive field allows us to take into account the range and roles of political, economic, social, and cultural relationships among the social actors.

Discursive fields encompass many elements, including social relationships, institutions, languages, conceptual frameworks, themes, discourses, and technological, economic, social, cultural, and political systems, all with histories. Sandra Harding (1988) refers to all the elements in the discursive field as "discursive resources."

For Foucault, discursive fields and their elements are structured in and through power relationships, and this is a critically important point. Modern power relations, he argues, are neither centralized nor localized but permeate the social world, resonating, coalescing, and strengthening each other (Foucault 1995, 26–31, 89–101; see also Foucault 1990, 92–

95). Power is not a commodity or possession in modernity but an exercise, a relationship, a practice that engenders resistance as it operates (Foucault 1990, 95–96). Knowledge and power are organically bound. "Power produces knowledge ... [they] directly imply one another, there is no power relation without the correlative constitution of a field of knowledge, nor any knowledge that does not presuppose and constitute at the same time power relations" (Foucault 1995, 27).

For Foucault, discourses cement the relationship between power and knowledge. But discourses are not, as the term is sometimes used, seamless, finished packages. Rather discourses are fragments, incomplete and in the process of being organized and reorganized in and through relations of power (Foucault 1990, 100–101). Knowledge is always constructed in specific configurations of historical, cultural, economic, and political relationships.

My analysis maintains a sense of fidelity to the notion of an unfinished, fragmented, contingent discursive field structured in and through relations of power. This point of analytical departure has ramifications for (1) the observation that the exercise of social control over the construction of knowledge is never a completely closed exercise, (2) considering the role of a plurality of voices in the creation of democratic practices in mass-mediated societies, and (3) the kinds of compromises and solutions to environmental conflicts, such as those over salmon and water resources, that we can both define and implement.

Economic equity and conservation are the two organizing discursive themes in the U.S.-Canadian Pacific Salmon Treaty. The text of the treaty cites conservation, rational management, and the promotion of optimum production as the principal interests of both Canada and the United States. The theme of economic equity is grounded in the notion that "states in whose waters salmon originate have the primary interest in and responsibility for such stocks" (Treaties and Other International Acts [TIAS] 11091 1985, 4). As such, a state has a right to "benefits equivalent to the production of salmon originating in its waters" (TIAS 11091 1985, Article III, 1b, 7). The Pacific Salmon Treaty aims to alleviate and redress the issue of excessive interceptions[4] through carefully defined and agreed upon procedures for establishing allocations (TIAS 11091 1985; TIAS 11778 1989; TIAS 11779 1988; TIAS 11839 1991).

The theme of conservation refers to the prevention of overfishing, the enhancement of fish habitats, and the issue of deciding and agreeing upon the escapements[5] necessary for the reproduction of the various stocks of salmon (TIAS 11091 1985).

Since the 1950s in both United States and Canada, the goals of increased economic efficiency and rationality in commercial fisheries production have significantly shaped fisheries policies. By the 1990s glaring declines in some economically high-profile fisheries stocks made biological reproduction, rather than harvesting, a more pressing issue, especially for U.S. fisheries managers.[6] However, economic rationality is still an influential goal in both U.S. and Canadian fisheries management practices for two reasons. First, the social goal of economic rationality in commercial fisheries production enjoyed a long history as the almost singular source of justification for fisheries policy. Second, economically pressed and productively constrained fishers exert their own political pressure to protect their livelihoods.

For the Canadian delegation, economic equity was a motivating factor for agreeing to the Pacific Salmon Treaty, but initially members of the delegation were willing to delay pressing their claims. By 1990, however, the Canadian delegation had begun to relentlessly press for economic equity between the nations in respect to commercial fisheries. They argued that the United States had repeatedly asked them to delay their equity claims. Canadians support conservation of the salmon stocks but argue that while they are conserving salmon, U.S. fishers reap the economic benefits through excessive interceptions of Canadian salmon (Pacific Salmon Commission 1991/92, 6–11; *The Pacific Salmon Treaty* 1997).[7] Especially at issue for the Canadians are the commercially valuable sockeye returning south to the Fraser River through Alaskan waters and salmon from the transboundary river systems returning to Canada through the Alaskan panhandle, especially in the Stikine, Taku, and Alsek Rivers (McDorman 1995, 491–497). Alaska's unique geographical siting allows fishers in Alaskan waters first access to returning Fraser River and transboundary river salmon stocks.

The U.S. delegation also subscribes to the economic equity principle but objects to the Canadian arguments about prioritizing equity in the negotiations. The U.S. delegation reclaims the category of equity by rejecting suggestions that a single economic value can, in practice, be computed for such widely different stocks and fisheries. They also assert that many of the nonmonetary values (symbolic, etc.) embedded in salmon cannot be adequately accounted for in purely monetary terms. And finally, the U.S. delegation appeals to conservation of the stocks and issues of possible extinction as a higher good than the economic equity principle (Pacific Salmon Commission 1991/92, 13–19). Especially at issue for the U.S. delegation is the conservation of depleted chinook and coho stocks from the Columbia and Snake River systems, which are targeted by Canadian fishers (Jensen 1986). The Canadians object to the United States invoking conservation in this manner and remain resolute in pressing for equity. They have repeatedly offered to enter into binding arbitration. The U.S. delegation steadfastly refuses. In May 1997, Westneat reported in *The Seattle Times* that former New Zealand Ambassador Christopher Beeby, acting as a nonbinding arbiter, found in favor of the Canadian position.

The U.S. delegation's appeal to conservation resonates with most of the U.S. environmental groups who are working to save the now endangered Columbia and Snake River wild salmon runs. Washington has an urbanized and industrialized economy. British Columbia's economy is rural and dependent on natural resource extraction industries, especially timbering, and to a lesser extent commercial fishing. The Canadian delegation's position on economic equity resonates with its constituency, which is by and large more directly concerned about an economic livelihood dependent on commercial salmon fishing.

Difficulties in negotiating agreements under the Pacific Salmon Treaty also arise from the fact that the nation-states themselves are not monolithic homogeneous blocks of interests (Huppert 1995, 14; Jensen 1986; McDorman 1995, 493). The multitude of jurisdictional authorities over salmon stocks, especially in the United States, complicates treaty negotiations and implementation.

Competing social actors within the nation-states also deploy the discursive themes of economic rationality, rights to resources, conservation, and sovereignty to assert their positions and interests. Furthermore, social actors within the nation-states exercise varying degrees of influence and control over the outcomes of treaty negotiations. As Huppert

(1995, 14) points out, Alaskan congressmen control key senior political posts, including chair of the Natural Resources Committee in the House of Representatives. Other less well represented voices, including the various treaty tribes in the United States, have pressed their claims through the U.S. federal court system.[8] The 1974 *Boldt* decision and the 1995 *Rothstein* decision demonstrate how conflicts within the United States serve both to constrain and to enable cooperation between the two countries (Huppert 1995, 7, 11; Jensen 1986, 381, 385–386, 389–393, 398–399, nn. 98, 100).[9]

McDorman (1995, 490) argues that the equity and conservation principles are contradictory and that the inclusion of both in the Pacific Salmon Treaty allows fishers to deploy different bits of the treaty to support their claims. His argument can also be extended to the other groups of social actors interested in and struggling over treaty outcomes. It is important to emphasize, however, that the salmon wars between the United States and Canada cannot be easily reduced to the opposition between the themes of economic equity and conservation. For all of the focus on conservation, struggles over economic resources are central to the contentious nature of the transborder public debates.

Article III of the treaty cites both "the desirability in most cases of reducing interceptions" and "avoiding undue disruption of existing fisheries" (TIAS 11091 1985, Article III 3a and 3b, 7). Different interpretations of "disruption" foment contention among the various players. Alaskans argue that curtailing their southeastern salmon fisheries will disrupt an existing fishery (Stevens 1986). From the Canadians' perspective, the U.S. argument is contradictory. The United States refuses to reduce its interceptions because of the threat of disrupting an existing fishery; however, it also insists that the Canadians reduce their interceptions for the sake of conservation and in spite of the disruption such an action would cause to Canadian production (Pacific Salmon Commission 1991/92, 8). Fundamental assumptions about economic rights to resources shape social expectations and political arguments over who gets to utilize and benefit from the income stream generated by salmon resources.

A third theme, after economic equity and conservation, that surfaces in the salmon wars is that of nationalism. Whereas conservation and equity are spelled out in the text of the treaty, the treaty process itself affirms the nation-state system, reinforcing the authority of the nation-state by reconfirming modernist notions about state sovereignty and by legitimating nation-based management regimes and national proprietary claims on salmon stocks.

Nationalism as a theme is prominently and frequently displayed in the mainstream media representations of the conflicts. Mainstream newspapers in both countries commonly employ militaristic metaphors to describe and analyze the salmon wars for their readers. This militaristic rhetoric is often tempered, however, by an additional nod to the nearly unguarded U.S.-Canadian border and the extensive trading partnership between the countries.[10]

The boundaries of many marine–fresh water systems do not neatly coincide with the territorial boundaries of modern nation-states. Governance regimes, however, remain anchored in a modernist system of nation-states. Currently, a number of authoritative biological arguments have been advanced as to why sound management practices of salmon fisheries stocks should be determined by the nation-state where the stocks originate (see Burke 1994, 153–154, 167–168; National Research Council 1996). These management rationales work to reinforce national claims and sovereignty—to a point. Contemporary fisheries management regimes deploy punitive sanctions as an enforcement vehicle, and this threat of coercive action should not be underestimated. However, implementing transboundary management decisions throughout the range of highly migratory fish depends rather heavily on the cooperation of harvesters and managers whose cooperation in the case of wild North American salmon stocks has been tenuous at best. The lack of enforceable cooperation has commonly presented serious challenges to proprietary claims on the stocks when they are migrating outside of the marine water belonging to the state in which the salmon originated. This conundrum is an example of contradictions inherent in glocalization processes.

Glen Clark, the premier of British Columbia in 1997, using various high-profile mass media campaigns, repeatedly exerted pressure on the United States and Canada in that year to resume the stalled Pacific Salmon Treaty talks. In one of those mass-media campaigns in early summer, Clark threatened to cancel the Canadian military's lease on the Vancouver Island land upon which the U.S.-Canadian joint-user, high-tech Nanoose Naval Base is located. Complex legal issues between the

Canadian federal and provincial governments arose from Clark's threats.[11] However, what is interesting about Clark's threat in the context of this discussion is the complex way it condenses the discursive themes of economic equity and nationalism in both a symbolic and material way.

Newspapers quickly seized the advantage of the discursive opening made by Clark to publish headlines such as "Torpedoes in a salmon war," which also featured a picture of the *U.S.S. Dolphin* submarine (see Anderson 1997, A1). Clark's move deployed jingoistic rhetoric and a potentially material attack on a military institution. Despite publicized opposition from the Canadian central government, he did not back down immediately. He claimed in the public forum that he did this to force the United States to address and resolve economic equity in salmon harvesting in a way that was meaningful for British Columbians. He targeted an international cooperative military institution in an economic war over natural resources.

Public arguments about salmon conservation and economic equity are steeped in nationalism. The public arguments are built on deeply anchored modernist assumptions about economic rationality, economic rights to resources, and governance regimes based on a system of sovereign nation-states. Examination of the public arguments also exposes deeply anchored culturally specific assumptions about the effectiveness and efficiency of using quantitative methods and technologically driven solutions to resolve environmental conflicts, especially conflicts that transcend the borders of nation-states. Both economic modeling and biological modeling rely heavily on quantification as a discursive source of authority. It is important to note that although social actors deploy these and other discursive resources to advance their positions and interests, the *available* discursive resources also significantly shape the positions and actions of the social actors.

Mass Media and Public Meanings

Public debates coalesce around the relationship between environmental conditions and economic production as social actors define and negotiate their positions with regard to environmental issues. Contemporary public debates are mediated through print presses, television, radio, and other electronic media. Social actors engaged in the international struggles over wild salmon resources include politicians, environmental groups, competing segments of the commercial fishing industry, and competing governments, including one provincial, four state, two national, and several First Nations and treaty tribe governments. All of the social actors involved in these debates utilize print media in some form to assert their positions in the public debates. Contemporary social actors look to the media for meanings and in turn use media to create public meanings.

Like Michel Foucault, Jürgen Habermas is also concerned with the construction of knowledge and social action in contemporary society, but his approach and assumptions are different from those of Foucault. Habermas places significantly less emphasis than Foucault on the constitutive role of power relations in contemporary society.[12] However, Habermas (1970) also posits two processes of knowledge creation and action that operate in two very different ways in contemporary Western society.

Purposive-rational decision making and instrumental action govern the first way of creating knowledge, according to Habermas. The emphasis here is on technical effectivity, rationality, and control. The second process for creating knowledge is the "action-orienting, self-understanding of social groups" (Habermas 1970, 52) in which, through debate and argument, shared knowledge and actions are created and negotiated. The procedures for arriving at shared knowledge legitimate the outcomes. Whereas the process of purposive-rational instrumental action colonizes public debates, practical or communicative action is normative and potentially emancipatory because of its democratic procedures (see also Held 1980, 249–295).

Purposive-rational action aims to reduce complex social problems to technical problems in a process that "not only justifies a particular form of class interest in domination, but also affects the very structure of human interests" (Held 1980, 254). The Pacific Salmon Treaty is an excellent example of an attempt to resolve conflicts in a highly fluid and often contentious transborder social system through apparently objective and socially neutral technical means.[13] The treaty is constructed around the idea that conflicts over allocations can be resolved through the application of mathematical techniques, including models that accurately predict the numbers of fish returning to spawn (Jensen 1986, 380–384, 392). The very models used to predict salmon runs are often sources of disagreement, however, among different governmental agencies responsible for generating databases concerning salmon resources. Tensions between agencies have increased since Jensen (1986) documented methodological problems engendered by the reliance on mathematical models in the late 1970s.

Arguments over the legitimacy of the various models used to predict salmon runs reveal deep conflicts about who should generate and control the knowledge about salmon resources. These political conflicts over models cannot be isolated from the social conflicts over economic rights to salmon resources and assertions of state sovereignty. Busch (1995) beautifully illustrates this point in her description of the scientific and political features of the incidents leading to the 1995 *Rothstein* decision to close Alaska's southeast chinook salmon fisheries early.

In 1995, Alaskan fisheries managers abruptly switched, in setting salmon harvest limits for Alaskan fisheries, from the salmon run prediction models employed by the bilateral Pacific Salmon Treaty Chinook Technical Committee to abundance-based modeling.[14] The decision immediately precipitated acrimonious debate within the United States as well as between the United States and Canada. Alaska's decision seemed to others to unfairly favor Alaskan fishers by allowing them to take larger catches, at least as long as everyone else was governed by regimes based on the prediction models (Busch 1995, 1507–1508; see also Huppert 1995, 15–16, for an explanation of abundance-based management).

Fisheries are human social systems with significant hydro-biological components, including fish and the marine–fresh water systems upon which these fish depend. The particular social system governed by the Pacific Salmon Treaty is exceedingly complex and contradictory, in part because salmon require a very extensive range of habitats that traverse the sovereign territorial boundaries of nation-states. In the 1995 incident, discursive assertions about economic rights and state sovereignty were integral to the seemingly socially neutral task of constructing quantitative models.

The creation of immense amounts of technical data is a discursive strategy that deserves even further consideration. Various U.S. federal and state agencies, Canada's Department of Fish and Oceans, and the two countries' joint international technical teams have all compiled and published thousands of pages of printed scientific information about salmon runs. The sheer quantity of this mostly quantitative data is a significant component of the publicly available printed material, which is also nearly impenetrable without technical expertise. Killingsworth and Palmer (1992) assert that the technical impenetrability of language in government reports excludes many people from the process of knowledge creation (see also chapter 11).

Assumptions about the social neutrality of quantitative modeling and the effectiveness and efficiency of technical solutions underlie the production of scientific knowledge about salmon runs. Fisheries management decisions are justified on the basis of those data. Quantification in both biological and economic modeling is a very authoritative discursive resource. These observations about the nature of data are not to argue that we do not need to understand the biological dimensions of natural resource systems; indeed we do need to understand the many components of natural resource systems, and technical analysis is one way to create knowledge about the biological components. However, Habermas (1970) suggests that when social conflicts are reduced to technical problems, other ways of representing and resolving the conflicts are marginalized. If resolutions do emerge, they do not take into account the plurality of interpretations and positions on the issues.

Government publications are one source of knowledge about salmon resources, and they are also one method of creating meaning in the public debates concerning those resources. Mainstream newspapers are another vehicle for creating public knowledge about the issues. Government agencies use newspapers to announce openings, catch quotas, and so on, as well as to present summaries of the quantitative data upon which their decisions rest. This suggests that mainstream presses serve as vehicles for informing the public and as venues for some accountability, however small, on the part of government agencies to the public. It also suggests that the public forum is never completely colonized by technocratic rationality and that technical solutions to issues and conflicts must be publicly defended.

Although a few formal forums have been organized to address the Pacific Salmon Treaty,[15] the news media are one of the most important vehicles for assertions of interests. I am certainly not suggesting that all social actors have equal access to creating knowledge in and through mass media, nor that their voices are somehow transparently represented when they do get press coverage. And most importantly, I point out that social actors utilizing the public forum are by and large limited to deploying the discursive elements already available.

Several recent volumes specifically examine the role of mass media in constructing environmental issues and environmental campaigns. The majority of the authors in these volumes examine the actions of the most powerful social actors such as media organizations, corporations, and governmental agencies, in environmental issues (Cantrill and Oravec 1996; Killingsworth and Palmer 1992; LaMay and Dennis 1991). Neuzil and Kovarik (1996) are an exception, in that they focus on less powerful social actors who mobilize press coverage to educate the public and garner support for their causes. My next two examples further explore the discursive strategies of social actors employing the print media to assert their positions.

Glen Clark's media campaigning (discussed above) during the summer of 1997 demonstrated his use of the discursive theme of nationalism. Another Clark media campaign boldly deployed the theme of conservation to combat a U.S. harvesting season on Stuart River sockeye salmon. On July 9, 1997, Clark ran a paid advertisement in *The Seattle Times*, the *Seattle Post-Intelligencer*, and *The Bellingham Herald*[16] in the form of an open letter addressed to the people of the state of Washington. It criticized the state's fishing interests for opening a season on the early Stuart sockeye salmon run "before the binational staff of the Pacific Salmon Commission could determine the true strength of the runs." It urged the people of Washington to pressure their state and federal governments to "conclude a Pacific Salmon Treaty before the fish themselves pay the price." The advertisement included a note labeling the letter as a paid advertisement, a map showing the locations of the early Stuart run, faint background images of swimming adult sockeye salmon, Clark's signature and official seal, and the names and addresses of the U.S. president, the Washington state governor, and the state's senators.

Regardless of Clark's motivations for waging such a provocative media campaign, his actions fomented an atmosphere of transnational public debate over fisheries governance regimes. The newspapers reported that a Bellingham, Washington, radio station refused to run the advertisement and instead turned Clark's ad campaign into the focus of a call-in talk show (see Beatty 1997a, 1997b; Bellingham Herald and Wire Services 1997; Simon 1997). The chair of the Northwest Indian Fisheries Commission and the chair of the Fish and Wildlife Commission jointly responded on the opinion page of the *Vancouver Sun* (see Frank and Pelly 1997).

Although international issues are ultimately negotiated between nation-states, they have an impact on local social actors. Canadian and American fishers whose livelihoods depend on the commercial harvesting of wild salmon have had to adjust to increasingly stringent regulations on salmon fishing. They have also been beleaguered by falling prices as farmed salmon has progressively displaced wild-capture salmon in world markets. With varying degrees of long-term impact, fishers create strategies and expand avenues for their own participation in the public debates over salmon harvest regulations. Periodically groups of fishers, as do many less powerful social groups, pool their resources, organize a united front, and force recognition of their position and interests.

In the summer of 1997, Canadian fishers and their supporters took their position to the streets, the waters, and the media. They staged their most effective international media event in July. Over the course of three days, between 150 and 200 Canadian fishing boats blockaded an Alaskan ferry in the port of Prince Rupert in British Columbia. A group of Canadian fishers staged a similar event in Prince Rupert in 1995 that garnered international print press coverage. The fishers in 1997 knew that such an action would get immediate press attention. Regional print coverage was constant during and immediately after the blockade. The image that most frequently appeared in both regional and national newspapers was an Associated Press wide-angle photograph showing numerous fishing boats encircling a comparatively massive ferry at the Prince Rupert dock. In late July several reports of another threatened blockade surfaced in the newspapers. In late August, newspapers reported that Alaska had filed a lawsuit against the blockaders and the government of Canada in Canadian Federal Court.[17]

Many people with opposing viewpoints expressed outrage, but nowhere in either Washington or British Columbia did reporters treat this event as a serious terrorist action. As the stories about the incident wound down, both *The Vancouver Sun* and especially *The Seattle Times* began to play up the shared interests of Canada and the United States.

Both U.S. and Canadian mainstream papers are owned and operated by large capitalist corporations that, at the very least, gate-keep the images and narratives available for consumption. The two largest newspapers in Washington, the *Seattle Times* and the *Seattle Post-Intelligencer*, are owned by the same corporation. Even a quick scan of the articles about the incident reveals that many of the narratives and the central image of the blockade were downloaded from the same wire service.

Gurevitch (1996, 219–221) argues that texts and especially visual images downloaded from sources such as wire services are actually filled with specifically local meanings that tap deeper, more abiding cultural myths. Globally available visuals and texts are both domesticated and worked to fit stable, locally familiar forms of narrative, although some myths may resonate more universally across cultures than do other myths (Gurevitch, Levy, and Roeh 1991, 206–208). The imagery of salmon wars echoes grander narratives about the military/industrial complex and nation-statehood. Images of organized, working fishing boats immobilizing a massive ferry echo cultural myths about the independent, entrepreneurial, hard-working spirit of fishers and about workers whose empowerment lies in organizing themselves against threats to their livelihoods.

Social actors pursue news coverage to raise awareness of their situation, knowing that other people watch the news, read the newspaper, and talk about events and issues. Access to the public debates, however, requires access to both material and symbolic resources that are unevenly distributed across the social landscape. Although many different voices work to make themselves heard in public debates through mass media, I emphasize that their positions are shaped by the *available* discursive resources. These constraints on democracy in public debates bear further consideration.

Conclusion

This chapter was written with the intention of using discourse analysis to challenge, critique, and contribute to the transformation of hegemonic ways of thinking about water resource management. It begins with the premise that the discursive themes and strategies deployed in public debates shape the ways that the conflicts over salmon and water resources are defined. When left unremarked and unquestioned, these themes and strategies can blind us to other kinds of compromises, alliances, and resolutions that we might create.

Although my method of discourse analysis is both critical and constructivist in its inclinations, the case under consideration is very much the product of its modernist patrimony. Although this book is about the growing challenges of transnational forces to modernity, mine remains a very important case study because the nation-state system is an entrenched social formation. Treaty processes are paramount vehicles in the contemporary world for negotiating and resolving international conflicts. Treaty processes are both the product, and in turn constitutive, of the modernist system of sovereign nation-states. The key discursive themes in this case study are also modernist in their claims and their logics. Discourse analysis brings to this arena a method for creating a deeper understanding of those claims and logics. That understanding can then serve as a basis for questioning and reworking both ideas about who can participate in the creation of knowledge and the very processes by which knowledge itself is created.

Nationalism is most obviously modernist, but claims of economic equity are also grounded in modernist notions. Economic equity depends on foundational assumptions about a system of private property rights that protects and legitimates claims of ownership. The very idea of private property reaches its apex of sophistication and influence in modernity in conjunction with an economic system of capitalism. Sovereignty in the modernist nation-state system is anchored in enforceable and enforced claims on territory and control over the resources in those territories, including water systems. The central, vivid, and frequently recurring representation of the issues, the salmon wars, describes a war over economic resources.

Capitalism is a productive system that encourages competition rather than cooperation, except perhaps for very self-interested cooperation. The discursive theme of nationalism serves as a vehicle for expressing the very competitive social relationships intrinsic to capitalist production.

Perhaps the theme of conservation goes the furthest in challenging modernist ways of understanding water conflicts, except that this challenge is not as straightforward as it seems. Conservation embraces many contradictory tendencies, and thus it may have emancipatory potential. But that potential will not be realized until the following points are addressed in a meaningful way.

Conservation is a very seductive theme because it suggests a higher good: the protection of resource systems on which everyone depends. Huppert (1995), for example, argues that the conservation measures fostered by the Pacific Salmon Treaty are a win-win situation because they will result in more salmon overall. However, his point obscures the issue of differential access to the material resources that everyone is supposed to work at fostering. For example, Alaska fishers have a geographic, and therefore economic, advantage in accessing the salmon stocks that British Columbians regard as essential to their livelihoods. The significance of differential access is not lost on the Canadians. Appeals to conservation can mask deep economic and social inequalities that need to be addressed as constitutive dimensions of conservation programs.

The discursive theme of conservation in Western countries like Canada and especially the United States is wedded to the idea that policies and management regimes are strongest and fairest when they are based on scientific expertise and findings. Conflicts over the role of science usually revolve around whether the science in question is good or bad. It is much harder to advance more fundamental questions about what role experts should play in resolving conflicts and about framing the issues at stake in such a way that technical findings provide answers to social conflicts (Yearly 1996, esp. 100–141). As Habermas has so astutely pointed out, the act of reducing social conflicts to technical problems excludes other kinds of knowledge and other voices from participating in the creation of governance regimes.

The relationship between conservation and technical solutions has two ramifications worthy of further consideration. First, the quantitative methodologies, such as modeling, used to support the discursive theme of conservation are not fundamentally socially neutral. The conflicts engendered by abundance-based modeling are an example of the fragility of discursive claims about the social neutrality of quantitative models.[18]

Second, technocratic rationality, which drives quantitative methodologies, also has played a seminal role in the development of the capitalist economic formation. Posing questions and seeking solutions through instrumental and utilitarian means has fostered mind-boggling scales of natural resource exploitation that need to be reconsidered if conservation is determined to be a socially significant goal. Furthermore, the very language that describes the variables to be measured in the models resonates with modern capitalist economic processes (e.g., resources, resource ownership, and stakeholders). These examples demonstrate both the constitutive nature of discursive practices and their embeddedness in the encompassing networks of economic and political relationships.[19]

Although globalizing processes challenge nation-state sovereignty, they also reconfirm the nation-state system. As Richard Perry (chapter 11) reminds us, both nation-states and modernity are recent inventions and very much the products of the narratives about them. And as he further reminds us, those narratives have both symbolic and material effects. The social system of nation-states has grown in tandem with the capitalist economic formation. That linkage needs to be further elaborated if we are to better understand the contradictions in processes of glocalization in the fluid arena of water resource conflicts and management.

Notes

1. The author thanks Joachim Blatter, Tom Boellstorff, Eve Darian-Smith, Helen Ingram, David Knowlton, Matt Mutchler, DeAnn Pendry, Richard Perry, Eric Poncelet, and Mayfair Yang, all of whom read and commented on versions of this chapter.

2. Canadians and Americans have shared and struggled over catches of wild North American Pacific salmon for nearly two centuries. After seventy-five years of international agreements that addressed portions but not all of their conflicts over salmon resources, the United States and Canada entered into the comprehensive Pacific Salmon Treaty in early 1985. However, the treaty was not in effect between 1994 and 1999. In 1999 yet another round of negotiations over the instrument were concluded.

3. Cantrill and Oravec 1996, Dryzek 1997, Hajer 1995, Milton 1996, and Williams and Matheny 1995 are all examples of recent publications in which discourse serves as an organizing theoretical construct for discussions about environmental and ecological issues.

4. "Interception" refers to "the harvesting of salmon originating in the water of one Party by a fishery of the other Party" (TIAS 11091 1985, 4).

5. "Escapement" refers to the idea that enough fish need to escape being harvested in a given season to spawn in sufficient numbers so as to sustain themselves both as a viably reproducing biological community and as a harvest commodity for the next season.

6. The National Research Council's 1996 book *Upstream* deals with Columbia River Basin salmon issues. An extensive team of biological and social science scholars with expertise on U.S. fisheries contributed to the volume, and their argument evinces a sea change in fisheries research positioning. They advocate completely subordinating harvesting goals to the goal of the biological reproduction of the salmon stocks because of an alarming loss of genetic variability and extensive depletion of stocks in the Columbia River Basin. They do not, however, place the blame solely or even primarily on fishers. Rather they take into account the whole range of factors that have an impact on salmon populations in the Columbia River Basin, including the loss of habitat through the development of an extensive hydroelectric industry (dams), timbering, urbanization, and management decisions.

7. The Canadian Consulate released a fourteen-minute video in the summer of 1997 in the United States that offers quantitative details supporting Canadian claims to equity compensation and excessive salmon interceptions by U.S. fishers. The video addresses equity by using U.S. biologists and conservationists almost exclusively as interviewees speaking about conservation. The video's central theme is economic equity, but it uses the rhetoric of conservation to assert Canadian equity claims.

8. The 1969 *Belloni* and 1974 *Boldt* decisions are very prominent examples. See Berg 1993 and Cohen 1986 for an account of these decisions and related subsequent rulings.

9. Huppert (1995) uses the 1974 *Boldt* and the 1995 *Rothstein* decisions as examples. Jensen (1986) offers an extended discussion about the influence of internal political pressures in the United States on the Pacific Salmon Treaty in the 1980s.

10. See Hume 1997 and Williams and Gilmore 1997 for extended examples of the use of militaristic metaphors. The phrases "salmon wars" and "fish war" describing this conflict also appear in scholarly articles, including Huppert 1995, 4; National Research Council 1996, 140; and Jensen 1986, 372, n. 18. Jensen comments on the recurring use of these terms.

11. Stories about Clark's threat to close Nanoose Navy Base include Anderson 1997; Associated Press 1997e; Beatty 1997c; Beatty and O'Neil 1997a, 1997b; Lee and Bramham 1997; O'Neil 1997; O'Neil and Beatty 1997; O'Neil with Hogben 1997; O'Neil and Ouston 1997; and Yaffe 1997. On July 14, 1997, *The Vancouver Sun* dedicated a third of its opinion page to readers' reactions to Clark's threat to close Nanoose. See Vancouver Sun 1997.

12. Habermas has been extensively criticized for systematically excluding whole categories of social actors from public-sphere participation. See Eley 1994, Fraser 1992, and Warner 1992.

13. Jensen (1986) repeatedly points out that Pacific Salmon Treaty negotiators have been highly attracted to complex technical methods for resolving conflicts over salmon resources.

14. The rationale behind the switch to abundance-based models is that they allow managers to adjust catch sizes during a salmon run as opposed to making a fixed preseason prediction. During larger than anticipated runs, managers can increase total allowable catches. During poor runs, they can curtail harvesting.

15. For example, a well-known Seattle environmental group and a tiny, fairly new transnational environmental group dedicated to addressing salmon biological sustainability as a transnational conservation issue organized such a forum in 1997.

16. Bellingham is a commercial fishing town located just south of the Canada-U.S. border.

17. Regional stories about the blockade include Associated Press 1997a, 1997b, 1997c, 1997d, 1997f; Associated Press and Reuters 1997; Ayers 1997; Fong and Pynn 1997; Hogben 1997a, 1997b; Hogben and Beatty 1997; Hogben and Bell 1997; Holt 1997; Keene 1997a, 1997b; Porter 1997; Seattle Post-Intelligencer News Service 1997a, 1997b, 1997c; and Sunde 1997. Examples of national coverage in the United States include a color rendition of the photo accompanied by a story about the struggles over salmon resources that appeared on the front of the business section of the *San Jose Mercury News* (Mercury News Wire Services 1997). In September the same photograph, in black and white, appeared in the *New York Times* with an extended story about regional conflicts in Cascadia (Eagan 1997).

18. Steven Yearly (1996) presents several transnational cases in which various parties challenge the scientific neutrality of the mathematical computer models being deployed. These cases include global climate modeling and global air standards models.

19. Michael Redclift (1993) critically analyzes the term "sustainable development" and its modernist roots in the scientific specialization of the disciplines, capitalist economic growth, and evolutionary theory. Ulrich Beck (1996) undertakes a similar critical examination of "risk societies."

References

Anderson, Ross. 1997. "Torpedoes in a Salmon War." *Seattle Times*, August 20, A1, A14.

Associated Press. 1997a. "Canadian Fishers Block Ferry." *Bellingham Herald*, July 20, A1.

Associated Press. 1997b. "Ferry Sits, Fishers Burn U.S. Flag." *Bellingham Herald*, July 21, A1, A2.

Associated Press. 1997c. "Senate Condemns Canada Blockade." *Bellingham Herald*, July 24, A1.

Associated Press. 1997d. "Prince Rupert Seeks Return of Ferry." *Bellingham Herald*, August 8, A3.

Associated Press. 1997e. "Salmon Mess Pits B.C. against Feds." *Bellingham Herald*, August 15, A3.

Associated Press. 1997f. "Alaska Goes After Canadians Who Blocked Ferry." *Bellingham Herald*, August 28, A1.

Associated Press and Reuters. 1997. "Canadian Court Orders Release of Alaska Ferry." *Seattle Times*, July 21, A1, A8.

Ayers, Scott. 1997. "Ferry Passengers Take Inconvenience in Stride." *Bellingham Herald*, July 23, A1, A2.

Beatty, Jim. 1997a. "U.S. Stations Ban Clark's Fish Ads." *Vancouver Sun*, July 10, A1, A2.

Beatty, Jim. 1997b. "U.S. Denounces Clark for Ad Campaign." *Vancouver Sun*, July 11, A7.

Beatty, Jim. 1997c. "Clark Hopes to Gain Support of Premiers in B.C.'s Salmon War." *Vancouver Sun*, August 6, A3.

Beatty, Jim, and Peter O'Neil. 1997a. "Premiers Leave Clark High and Dry in His Nanoose Stand over Salmon." *Vancouver Sun*, August 8, A1, A2.

Beatty, Jim, and Peter O'Neil. 1997b. "Nanoose Stays Open Pending Court Case." *Vancouver Sun*, August 15, A1, A21.

Beck, Ulrich. 1996. "World Risk Society as Cosmopolitan Society: Ecological Questions in a Framework of Manufactured Uncertainties." *Theory, Culture, and Society* 13, no. 4: 1–32.

Bellingham Herald and Wire Services. 1997. "B.C. Premier Roils the Salmon Waters." *Bellingham Herald*, July 11, A3.

Berg, Laura. 1993. "Let Them Do as They Have Promised." *Hastings West-Northwest Journal of Environmental Law and Policy* 3 (Fall): 7–17.

Burke, William T. 1994. *The New International Law of Fisheries: UNCLOS 1982 and Beyond.* Oxford: Clarendon Press.

Busch, Lisa. 1995. "Scientific Dispute at Center of Legal Battle over Salmon Catch." *Science* 269, no. 5230: 1507–1508.

Cantrill, James G., and Christine L. Oravec, eds. 1996. *The Symbolic Earth: Discourse and Our Creation of the Environment.* Lexington, Ky.: University Press of Kentucky.

Cohen, Fay. 1986. *Treaties on Trial.* Seattle: University of Washington Press.

Dryzek, John S. 1997. *The Politics of the Earth: Environmental Discourses.* Oxford: Oxford University Press.

Eagan, Timothy. 1997. "Salmon War in Northwest Spurs Wish for Good Fences." *New York Times*, September 12, A1, A14.

Eley, Geoff. 1994. "Nations, Publics and Political Cultures: Placing Habermas in the Nineteenth Century." In Nicholas Dirks, Sherry Ortner, and Geoff Eley (eds.), *Culture/Power/History*, pp. 297–335. Princeton, N.J.: Princeton University Press.

Fong, Petti, and Larry Pynn. 1997. "Alaskans Having a Hard Time Forgetting Ferry Blockade." *Weekend Sun*, September 20, A1, A20.

Foucault, Michel. 1972. *The Archaeology of Knowledge.* New York, London, and San Francisco: Harper Colophon.

Foucault, Michel. 1990. *The History of Sexuality. Volume 1: An Introduction.* New York: Vintage Books.

Foucault, Michel. 1995. *Discipline and Punish: The Birth of the Prison.* New York: Vintage Books.

Frank Jr., Billy, and Lisa Pelly. 1997. "U.S. View: Salmon Ad is B.C. Propaganda." *Vancouver Sun*, July 21, A11.

Fraser, Nancy. 1992. "Rethinking the Public Sphere: A Contribution to the Critique of Actually Existing Democracy." In Craig Calhoun (ed.), *Habermas and the Public Sphere*, pp. 109–142. Cambridge: MIT Press.

Gurevitch, Michael. 1996. "The Globalization of Electronic Journalism." In James Curran and Michael Gurevitch (eds.), *Mass Media and Society*, pp. 204–224. London and New York: Arnold, Hodder Headline Group.

Gurevitch, Michael, Mark R. Levy, and Itzhak Roeh. 1991. "The Global Newsroom: Convergences and Diversities in the Globalization of Television News." In Peter Dahlgren and Colin Sparks (eds.), *Communication and Citizenship: Journalism and the Public Sphere*, pp. 195–216. London and New York: Routledge.

Habermas, Jürgen. 1970. *Toward a Rational Society: Student Protest, Science, and Politics.* Boston: Beacon Press.

Hajer, Maartin. 1995. *The Politics of Environmental Discourse: Ecological Modernization and the Policy Process.* Oxford: Clarendon Press.

Harding, Sandra. 1998. *Is Science Multicultural? Postcolonialisms, Feminisms, and Epistemologies.* Bloomington and Indianapolis: Indiana University Press.

Held, David. 1980. *Introduction to Critical Theory: Horkheimer to Habermas.* Berkeley and Los Angeles: University of California Press.

Hogben, David. 1997a. "B.C. Fish Protesters Urged to Cool Down." *Vancouver Sun*, July 30, A1.

Hogben, David. 1997b. "Blocked Ferry Was Due to Be Pulled from Run." *Vancouver Sun*, August 1, A1.

Hogben, David, and Jim Beatty. 1997. "Alaska Indicates Ferries to B.C. Will Resume Soon." *Vancouver Sun*, July 29, A3.

Hogben, David, and Stewart Bell. 1997. "B.C. Fishers Defiant in Spite of Order to End Blockade of Ferry." *Vancouver Sun*, July 21, A1, A2.

Holt, Gordy. 1997. "Politics May Gut Fishing Friendships." *Seattle Post-Intelligencer*, July 21, A1, A6.

Hume, Steven. 1997. "The Great 1997 Salmon Showdown Made Easy." *Vancouver Sun*, July 8, B8.

Huppert, Daniel D. 1995. "Why the Pacific Salmon Treaty Failed to End the Salmon Wars." *School of Marine Affairs, University of Washington 95*, no. 1: 1–27.

Jensen, Thomas C. 1986. "The United States–Pacific Salmon Interception Treaty: An Historical and Legal Overview." *Environmental Law 16*, no. 3: 363–422.

Keene, Linda. 1997a. "Ferry Freed, but Fish War Far from Over." *Seattle Times*, July 22, A1, A12.

Keene, Linda. 1997b. "Fish Wars are Talk of the Town." *Seattle Times*, July 23, B1, B2.

Killingsworth, Jimmie, and Jacqueline S. Palmer. 1992. *Ecospeak: Rhetoric and Environmental Politics in America.* Carbondale: Southern Illinois University Press.

LaMay, Craig L., and Everette E. Dennis. 1991. *Media and the Environment.* Washington, D.C., Corvelo, California: Island Press.

Lee, Jeff, and Daphne Bramham. 1997. "B.C. Can't Cancel Nanoose Lease, Ottawa Says." *Vancouver Sun*, August 6, A3.

McDorman, Ted C. 1995. "The West Coast Salmon Dispute: A Canadian View of the Breakdown of the 1985 Treaty and the Transit License Measure." *International and Comparative Law Journal 17* (April): 477–505.

Mercury News Wire Services. 1997. "Ferry Blockade in U.S.-Canada Fishing Dispute Is Ended." *San Jose Mercury News*, July 2, C1.

Milton, Kay. 1996. *Environmentalism and Cultural Theory: Exploring the Role of Anthropology in Environmental Discourse*. London, New York: Routledge.

National Research Council (Commission on Life Sciences, Board of Environmental Studies and Toxicology, Committee on Protection and Management of Pacific Northwest Anadromous Salmonids) 1996. *Upstream: Salmon and Society in the Pacific Northwest*. Washington, D.C.: National Academy Press.

Neuzil, Mark, and Bill Kovarik. 1996. *Mass Media and Environmental Conflict: America's Green Crusades*. Thousand Oaks, Calif., and London: SAGE.

O'Neil, Peter. 1997. "Fish War: Ottawa Eyes Nanoose Seabed." *Vancouver Sun*, July 9, A1, A4.

O'Neil, Peter, and Jim Beatty. 1997. "Clark 'Sought Way Out' of Nanoose Threat." *Vancouver Sun*, August 7, A1, A4.

O'Neil, Peter, with David Hogben. 1997. "Nanoose Action Dangerous, B.C. Business Groups Claim." *Vancouver Sun*, August 14, A5.

O'Neil, Peter, and Rick Ouston. 1997. "Ottawa Weighs Court Challenge over Nanoose." *Vancouver Sun*, August 13, A1.

Pacific Salmon Commission. 1991/92. *Pacific Salmon Commission Seventh Annual Report*. Vancouver, British Columbia: Pacific Salmon Commission.

The Pacific Salmon Treaty. 1997. Canadian Consulate General and Terry Barksdale Media Productions, Seattle, Washington. 14 min. VHS videocassette.

Porter, Mark. 1997. "Fishers End 3-day Ferry Blockade." *Bellingham Herald*, July 22, A1.

Redclift, Michael. 1993. "Sustainable Development: Needs, Values, and Rights." *Environmental Values* 2: 3–20.

Seattle Post-Intelligencer News Service. 1997a. "Canada Court Order Ferry Freed." *Seattle Post-Intelligencer*, July 21, A1, A6.

Seattle Post-Intelligencer News Service. 1997b. "Blockade Ends; Ferry Sails." *Seattle Post-Intelligencer*, July 22, A1, A7.

Seattle Post-Intelligencer News Service. 1997c. "Alaska Ferry May Again Face Blockade." *Bellingham Herald*, July 29, A1.

Simon, Jim. 1997. "Ads Latest Weapon in Fish War." *Seattle Times*, July 10, B1, B2.

Stevens, Ted. 1986. "United States–Canada Salmon Treaty Negotiations: The Alaskan Perspective." *Environmental Law* 16, no. 3: 423–430.

Sunde, Scott. 1997. "Alaska Ferry Braces for Another Blockade by Angry Fishermen." *Seattle Post-Intelligencer*, July 29, A1, A5.

Treaties and Other International Acts Series (TIAS) 11091. 1985. Treaty between the United States of America and Canada: Fisheries, Pacific Salmon. Pittsburgh, Pa.: U.S. Government Printing Office.

Treaties and Other International Acts Series (TIAS) 11779. 1988. Agreement between the United States of America and Canada: Fisheries, Pacific Salmon. Pittsburgh, Pa.: U.S. Government Printing Office.

Treaties and Other International Acts Series (TIAS) 11778. 1989. Agreement between the United States of America and Canada: Fisheries, Pacific Salmon. Pittsburgh, Pa.: U.S. Government Printing Office.

Treaties and Other International Acts Series (TIAS) 11839. 1991. Agreement between the United States of America and Canada: Fisheries, Pacific Salmon. Pittsburgh, Pa.: U.S. Government Printing Office.

Vancouver Sun. 1997. Editorial. July 14, A11.

Warner, Michael. 1992. "The Mass Public and the Mass Subject." In Craig Calhoun (ed.), *Habermas and the Public Sphere*, pp. 377–401. Cambridge: MIT Press.

Westneat, Danny. 1997. "Secret Fishing Report Sides with Canada." *Seattle Times*, May 31, A1, A8.

Williams, Bruce A., and Albert R. Matheny. 1995. *Democracy, Dialogue, and Environmental Disputes: The Contested Languages of Social Regulation.* New Haven and London: Yale University Press.

Williams, Marla, and Susan Gilmore. 1997. "Sockeye Catch off Alaska Is Key to U.S.-Canada Salmon Dispute." *Seattle Times*, July 22, A10.

Yaffe, Barbara. 1997. "U.S. Views Nanoose Bay Closure Threat as a Scuffle between B.C., Ottawa." *Vancouver Sun*, July 11, A3.

Yearly, Steven. 1996. *Sociology, Environmentalism, Globalization: Reinventing the Globe.* London and Thousand Oaks, Calif.: SAGE Publications.

7

Discourses and Water in the U.S.-Mexico Border Region

Pamela M. Doughman

Based on their analysis of efforts to site new hazardous waste management facilities in the United States, Williams and Matheny (1995) argue that there are three influential discourses in environmental regulatory policy: managerial, pluralist, and communitarian. Williams and Matheny suggest that the biggest difficulties in environmental infrastructure projects are caused by the failure to listen to the communitarian discourse.

In this chapter, I discuss the dominance of the managerial discourse at the International Boundary and Water Commission (IBWC) and the pluralist and communitarian challenges to the IBWC discourse. The creation of two new institutions, the Border Environment Cooperation Commission (BECC) and the North American Development Bank (NADBank) was justified by scholars and environmentalists as a policy response to challenges to the IBWC (Texas Center for Policy Studies 1992, Ingram and White 1993). I use discourse analysis to determine how far the BECC-NADBank discourse goes toward addressing the pluralist and communitarian critique. In my analysis I use laws, rules, and interviews to analyze the institutional discourse of the BECC and NADBank.[1] I also use published statements of professional, academic, activist, and government critics to identify key arguments and problem definitions that influence the discursive context in which institutions and implementing agents have acted.

The way water is viewed or framed by institutions involved in financing water infrastructure affects the range of alternatives they accept as possible courses of action. The ideational or discursive framework within which an institution operates is delineated by the laws, rules, and rationalizations establishing and established by the institution itself. These items show the language that the institution uses to legitimate and advance its aims.

Individuals involved in implementing an institutional framework affect its manifestation through their interpretation and inclinations. Target groups also influence implementation through the reception they give to institutional efforts to affect their well being or change their behavior. To understand the way institutional discourse is translated through the implementation process, it is necessary to understand the frame of reference that implementing agents and members of target groups employ.

As described in chapter 2, modern meanings of water focus on water as property, water as a product, and water as a commodity. For postmodernists, water is not seen as a material thing to be owned, made, or purchased. Rather, it is often an idea, a symbol, or even a virtual reality. Building on the ideas of chapter 2, García-Acevedo suggests in chapter 3 another postmodern meaning of water that is evidenced in the Imperial and Mexicali Valleys of California and Mexico: She notes that water can be understood as a source of community building.

In this chapter, I return to the U.S.-Mexican border and Mexicali to see how the meaning of water for the BECC and NADBank changes and shifts between modern and postmodern ontologies. The BECC and NADBank articulate clearly modern views of water by stressing its manufacturability and its commensurability. At the same time, they engage in a postmodern discourse that defines the meaning of water as a source of community building. This chapter contends that although these institutions broaden the terms of debate and provide avenues for increased dialogue within and among interested communities, the BECC and NADBank remain committed to a largely modern ontology and a primarily managerialist discourse.

The Managerial Discourse and the IBWC

The managerial mode of discourse stresses the importance of finding solutions to societal problems through scientific inquiry. Its standard of justice is utilitarian efficiency, defined as the conditions under which the total benefits of a policy to society outweigh the total costs. The role of a "good" citizen is to accept the "superior knowledge" of experts who rely on science to produce information (Fischer 1990; Schneider and Ingram 1997). Managerialism was advanced by early-twentieth-century Progressives in the United States to guard against abuse of power by special interests (see Williams and Matheny 1995, 11–19). With their emphasis on efficiency and expert knowledge, large-scale bureaucracies are an organizational form highly compatible with managerialism.

Since 1944, the U.S. and Mexican sectors of the IBWC have been the primary conduit for decisions and funding on U.S.-Mexico water issues. The IBWC has employed product-type engineering solutions to water quality problems and a formal bilateral approach to decision making. According to the terms of the U.S.-Mexico water rights treaty of 1944,[2] which assigned the IBWC its water-related duties, top decision-making positions in the IBWC must be assigned to engineers. An engineer must head the IBWC, and both the U.S. and the Mexican sections must contain two principal engineers.

The IBWC has lost credibility among concerned border residents, academics, and environmentalists because it has fallen short of achieving its own goals (Ingram and White 1993). The IBWC has been criticized from within the managerial discourse for not delivering water quality or wastewater service in a timely and effective manner. In chapter 3, García-Acevedo described the IBWC's efforts to address the high salinity of Mexico's allocation of water from the Colorado River. Despite the IBWC's efforts, however, local groups in Mexicali and others argue that periodic spikes in salinity of the Colorado River water received from the United States indicate that the salinity problem has not yet been resolved (Int. 85, confidential interview, March 1999).

In addition to dealing with water salinity problems, the IBWC is also expected to manage untreated domestic and industrial wastewater in the border region between the United States and Mexico. Over time it has not met this expectation to the satisfaction of citizens living within its sphere of influence. Ingram and White (1993) chronicled the inadequacy of IBWC response to sewage problems in Nogales, Arizona, and Nogales, Sonora. Since 1946, according to Ingram and White, IBWC efforts have failed to keep up with the wastewater treatment needs of these two cities:

By 1958, just seven years after the first international wastewater treatment facility was completed, the plant design capacity was being continuously exceeded and raw sewage was being bypassed into the Nogales Wash.... An agreement between Mexico and the United States was finally reached in 1967.... The new facility was completed in 1971. By 1976 the need for further expansion was already evident, and once again planning, design, and negotiations for enlargement of the facility were lengthy and tortuous. In 1986 a formal proposal was made to the Mexican Commissioner from the United States Section suggesting that both countries expand the International Wastewater Treatment Plant.... Expansion of the facility began in 1989 ... and the plant's design capacity of 15.75 million gallons per day (mgd) was exceeded during periods of peak usage even before the facility became fully operational. In addition, the treated effluent from the plant was not meeting Arizona water quality standards due to high levels of mercury, cyanide, and other pollutants, presumably due to the lack of industrial pretreatment of wastes. (Ingram and White 1993, 162–164)

The situation in Nogales is not unique. Continuing population growth, urbanization, and industrialization in many areas along the border have created a condition that the American Medical Association characterized in 1990 as a "virtual cesspool and breeding ground for infectious disease" (Hendee 1990). The American Medical Association report, and the media attention it received, helped to raise health-related water problems along the U.S.-Mexican border to the national level of policy debate on environmental concerns associated with the North American Free Trade Agreement (NAFTA) (Audley 1997, 50).

The characterization of water at the border as a threat to human health was further reified by the U.S. Environmental Protection Agency (EPA) and its counterpart in Mexico, the Secretaria de Desarrollo Urbano y Ecologia (SEDUE) (Secretariat of Urban Development and Ecology), established in 1982. In 1991 EPA and SEDUE described border water quality conditions as "a strain on the Border Area's infrastructure" (EPA-SEDUE 1991, 1-3):

Water quality in the border area is threatened by limited sewage treatment and collection facilities, untreated or inadequately treated industrial effluents, and improperly handled hazardous wastes. Inadequately treated wastewater flows into the Rio Bravo/Rio Grande, Colorado and other border area rivers, causing conditions that present a significant health risk and resulting in drinking water safety being a public concern.

In addition to concerns about its handling of surface water issues, the IBWC has also been criticized for failing to address long-standing concerns that precious groundwater reserves in the border area are being depleted and contaminated. In 1973 the IBWC was tasked by its Minute 242 to negotiate an international agreement on groundwater.

The agreement referred to in the Minute has not yet materialized. In 1984, Utton reported that El Paso and Ciudad Juarez were rapidly drawing down their aquifers and that Arizona was overdrawing its aquifers by 2.5 million acre-feet per year (Utton 1984, 14–15). In 1995, the EPA held meetings with citizens in eleven U.S. border cities to address environmental issues. Citizens attending these meetings in six of the eleven cities raised concerns regarding pollution of groundwater: San Diego, California; Nogales and Douglas, Arizona; Las Cruces, New Mexico; and El Paso and McAllen, Texas (EPA 1996).

Challenges from the Pluralist and Communitarian Discourses

Augmenting criticisms from within the managerial perspective, the IBWC has also been criticized in the pluralist and communitarian discourses. In the pluralist discourse, the role of public policy in society is to represent interests, resolve conflicts, allocate goods and services, and maintain system stability. Policy change occurs through considered compromises among competing perspectives (Lindblom 1979). No single actor or institution controls whether government acts on an issue: Rather, action can result from a confluence of workable solutions, accepted problem definitions,[3] and a supportive political context coordinated by policy entrepreneurs (Kingdon 1984). Standards of justice center on due process, the rule of law, and protection of individual rights and liberty. The role of a "good" citizen is to vote, communicate interests to political leaders, mobilize when interests are threatened, and support the democratic structure of government (Schneider and Ingram 1997).

The communitarian discourse emphasizes discursive democracy, reduction of oppression, and empowerment of local people (Barber 1998). Communitarianism insists that individuals do not exist in isolation; they are rooted in community. As Amitai Etzioni (1983) notes, "the

individual and the community make each other and require each other"
(Etzioni 1983, 25, as cited in Daly and Cobb 1994, 18). In the communitarian discourse, standards of justice center on principles of equality,
equity, or need. The role of a "good" citizen is to engage in discourse,
look for areas of agreement among people, actively strive for consensus,
and participate directly in local government (e.g., town hall meetings).
Furthermore, the local context is critical to understanding political and
social conditions and solutions (Williams and Matheny 1995; Schneider
and Ingram 1997).

During the negotiations over NAFTA, concern over U.S.-Mexico border water conditions prompted calls for changes, including the creation
of new institutions, to better manage the border crisis. For example, the
Texas Center for Policy Studies (TCPS), a nongovernmental organization
concerned with environmental and community health in the border region, criticized the IBWC for failing to follow a process of deliberation
and decision making that is open to the public:

IBWC is a fairly closed agency, viewing its functions as primarily diplomatic.
There is no public participation in Commission decision-making, with the exception of Environmental Impact Statements for major construction projects.
Commission meetings are closed to the public. The U.S. Section of the IBWC has
acknowledged that it is subject to the Freedom of Information Act. However,
it treats almost all information from Mexico as confidential and it has an
antiquated record-keeping system that makes it difficult for the public to get
complete responses to FOIA [Freedom of Information Act] requests. (TCPS
1992, 3-1)

The above passage is consistent with standards of justice important to
the pluralist discourse: due process, rule of law, and individual rights and
liberty. Here the TCPS is criticizing the IBWC for failing to follow established procedures intended to protect citizens' right to know. In addition,
it portrays avenues for citizens to register concerns with the IBWC as circumscribed by the limited opportunities for public attendance at IBWC
meetings.

In a somewhat resigned tone that is obliquely supportive of communitarian views of governance, Sanchez (1993) argues that increased public
participation in IBWC decision making seems "unavoidable" but likely
to produce positive effects. Sanchez asserts that public participation can
provide a forum for organized groups to develop a unified set of inter-

ests. He also states that public participation can provide a mechanism for border communities to "solve or improve their already existing environmental problems ... [and] to defend their needs and interests against the federal agencies, including the IBWC" (296). Sanchez's view supports the communitarian emphasis on the role of public forums for discussion and development of common goals within a community. By articulating a need to defend local interests against the federal government, Sanchez makes an argument similar to that made by TCPS: that efforts should be made to ensure that local interests in the border region are not left out of federal-level decision making.

The TCPS also employed arguments consistent with the communitarian discourse in its critique of the IBWC. The TCPS suggested the following rationale and course of action to Congress:

Both sides of the U.S./Mexico border are in dire need of major infrastructure improvements. In the area of sewage alone, most of the large Mexican border cities are currently without treatment systems.... [T]here are not enough federal funds to go around and most areas of the border lack the political clout necessary to secure adequate federal revenues for their needs.... Options for Congress:... Create North American Regional Development Bank and Assistance Fund.... The structure and governance of the Bank would have to be carefully crafted to be responsive to border region communities, while still maintaining financial accountability. (TCPS 1992, 5-1, 5-2)

This passage reflects the communitarian view that justice is based on equity and need. TCPS reported that the distribution of available moneys for wastewater projects was concentrated in a few powerful cities (i.e., Tijuana, Mexicali, Nogales, and Nuevo Laredo) rather than spread throughout the border region. Cities with needs (i.e., Matamoros) similar to the needs of those receiving funding were portrayed as being treated inequitably. The fact that other cities in the border region did not have sufficient political influence to win funding from the federal government for wastewater infrastructure was not seen as a function of an acceptable or fair process. Rather, the existing process left communities with unmet health needs and for that reason should be altered.

The passage also specifies a requirement that the bank "be responsive to border region communities." This too reflects the communitarian discourse, as it focuses on the local rather than the national as the optimal context for understanding and resolving wastewater treatment needs. Through responsiveness to border communities, TCPS implied, the new bank would advance the communitarian goal of empowering local people.

Response to Discursive Challenges

Drawing upon managerial, pluralist, and communitarian discourses, challenges to the IBWC during the NAFTA debate focused on water in the U.S.-Mexican border environment as a health, pollution, and infrastructure problem. Scarcity of water was acknowledged, but only product-type solutions (e.g., reuse, reclamation, and conservation) were emphasized. This view of the problem is reflected in the institutions that resulted from the NAFTA debate.

The BECC and NADBank were created to respond to the view that there is a health-threatening wastewater infrastructure deficit in the border region. That this is the primary function of these institutions is made clear in the agreement by which they were created:

> The purpose of the Commission shall be to help preserve, protect and enhance the environment of the border region in order to advance the well-being of the people of the United States and Mexico.... In carrying out this purpose, the Commission shall cooperate as appropriate with the North American Development Bank and other national and international institutions, and with private sources supplying investment capital for environmental infrastructure projects in the border region."[4]

The development of wastewater infrastructure is linked in the BECC and NADBank discourse to the concept of sustainable development. The United States and Mexico decided to create the NADBank and BECC, in part, because they were "convinced of the importance of the conservation, protection and enhancement of their environments and the essential role of cooperation in these areas in achieving sustainable development for the well-being of present and future generations."[5]

As all NADBank projects must obtain BECC certification, the BECC's approach to sustainable development applies for both institutions. The BECC has established both required and optional certification criteria for sustainable development. To receive certification from the BECC, projects must include statements and materials supporting (1) the

BECC's sustainable development definition and principles, (2) institutional and human capacity building, (3) conformance with applicable local and regional conservation and development plans, (4) natural resource conservation, and (5) community development.

The optional "high sustainability recognition" process is BECC's more complete statement on sustainable development. Characteristics of BECC's view of high sustainability include ecosystem-based planning, conservation of resource utilization and minimization of waste production, equitable distribution of costs and benefits, and a significant role for local people in "construction, operation, and maintenance of a project over its life cycle" (BECC 1996, 39).

The existence of two separate standards, one mandatory and one optional, suggests that the BECC perceives the optional standard to be unobtainable by at least some of its client base. Whether because of urgency of project completion, lack of interest, expense, or inability to provide the requested information for high sustainability, during the first two years of the criteria's existence, none of the proponents of the projects approved by BECC elected to complete the high-sustainability approval process.[6]

The BECC's sustainability criteria require articulation of support for environmental values in project proposals. The optional high sustainability criteria request information that indicates whether a project "encourages urban and industrial development as well as what impact it might have on the neighboring regions." A description of the "source of water for water supply, recharge area, monitoring and prevention of surface and/or groundwater contamination, etc." is also requested (BECC 1996, 39). Together with the requirement that proposed development projects provide for present and future generations' needs, this information could be used to launch a discussion of the limits to growth of cities in the fragile desert environment of the U.S.-Mexico border. The fact that such discussions are not characteristic of BECC and NADBank projects can be best understood through an analysis of the meanings of water that these institutions embrace.

According to BECC/NADBank discourse, new water infrastructure projects delivering cleaner product can mitigate the environmental problems of the U.S.-Mexico border. Furthermore, because of the transnational nature of the waterways in the area, local water pollution problems require special binational institutions that represent layers of interests and promote intergovernmental cooperation and communication.

The wastewater problems in the city of Mexicali in the Mexican state of Baja California illustrates some of the ramifications of viewing water as a product. As García-Acevedo described in chapter 3, manipulation of the Colorado River, including the formation of the New River, has had a profound effect on the people of Mexicali. Today the New River serves as the main conduit of wastewater out of the Mexicali Valley. The current wastewater treatment plant in Mexicali is unable to handle the volume of sewage and industrial wastes generated in the city. Thus, the New River drains raw sewage and untreated industrial wastes across the U.S.-Mexico border into the Salton Sea.

The IBWC and the Centro Interdisciplinario del Lenguaje y Aprendizaje (CILA) have been working on a solution to the wastewater problems of Mexicali. In support of this effort, the BECC has certified the two organizations' plan to produce cleaner water in the New River, and the NADBank is planning to contribute to the financing of this project. Construction of the $50 million Mexicali project is made possible by collaboration of a number of governmental entities: the IBWC, the CILA, the BECC, the NADBank, and federal, state, and local agencies in United States and Mexico. The IBWC and the NADBank (through the EPA-sponsored Border Environment Infrastructure Fund (BEIF) fund) will pay for 55% of the cost of the project, and the Mexican government will pay for the remaining 45%. In contrast to the existing wastewater treatment system, the current project has a guaranteed source of money for operation and maintenance: Mexico and the United States have indicated that they will establish two reserve funds through the IBWC/CILA for this purpose.[7]

The documentation promoting the Mexicali plan is rooted in the managerial worldview. The following excerpt illustrates the technocratic considerations used to justify selection of the proposed site for the new Mexicali II water treatment facility:

From these basic treatment processes the natural lagoons alternative was selected because of its lower costs.... In the same regard, a technical-economical analysis has been made in order to select the site of the Mexicali II wastewater treatment plant. (State Public Services Commission of Mexicali 1997, 14)

In this description, the planning process is framed as being based on cost and other utilitarian considerations. Although both institutions include cost considerations in project evaluation, the BECC-NADBank discourse differs from that of the IBWC as to who should pay for wastewater infrastructure. Whereas IBWC projects are paid for by federal dollars, a pillar of NADBank policy is that local communities should shoulder operation and maintenance costs of projects it finances:

In carrying out its mission, the NADB[ank] recognizes that both the United States and Mexico are seeking to achieve a gradual transition from projects that are fully subsidized by grants and government budget allocations to projects that are fiscally sound, financed under competitive market conditions, and serviced by user fees or other revenue. (NADBank 1998b, 1)

Shifting the burden of paying for wastewater infrastructure from the federal to the local level along the border was viewed as a revolutionary idea by policymakers in Washington. It was also seen by these policymakers as a way to give local communities more control over whether the project remained functional over time:

During the NAFTA debate ... [there was] a call for empowering local communities to deal across the boundary on these problems.... [T]hat meant that the burden of paying for these facilities would be shifted from the U.S. federal government to the local communities.... If people had a stake in [the wastewater treatment facilities] on both sides of the boundary,... they would take responsibility for making sure they were properly operated and maintained and the funds were put in them. So, this was really a revolutionary concept in many ways. It was quite a change from the way things had been done in the past. (Int. 16, confidential interview, August 1997)

Asking local communities to shoulder the cost of wastewater infrastructure was also seen by high-ranking policymakers in Washington as a way to extend the longevity of the NADBank. The NADBank was designed to leverage a one-time infusion of capital stock of $3 billion ($0.45 billion in capital shares; $2.55 billion in callable shares)[8] to build environmental projects addressing water issues. The NADBank was intended to be independent of the need for periodic replenishment:

We wanted to ... create an institution that was sustainable by itself. Which meant that it couldn't be necessarily a grant making institution, because that would require constant replenishment of the fund. So we determined that we had to kind of fit a development bank role, but . . . looking at development banks, we decided that it had to address some of the criticisms that existing development banks were under. And I think that . . . quite a number of people felt that . . . to create an institution and the policies around that institution, we had to learn from the mistakes of others. And that was part of the intent: to ensure more transparency, to ensure a more democratic way of determining just how development was going to take place. (Int. 9, confidential interview, August 1997)

The BECC and NADBank articulate a third meaning of water that responds more clearly to criticisms of the pre-NAFTA discourse. In addition to seeing water as a product and a commodity, the BECC and NADBank view water as a source of community building. This third meaning of water embraced by the BECC and NADBank leads the two organizations to promote public discussion of water issues.

The project design and capacity-building mechanisms advocated by the NADBank and BECC in their promotional literature reflect substantial respect for private citizens' opinions, abilities, and potential. For example:

Without the active support and participation of all those involved in environmental infrastructure development, the needs of many communities on both sides of the border could not be effectively served. The NADB[ank] and BECC firmly believe that by working together and combining resources, substantial progress can be made toward creating a cleaner and healthier environment for current border residents, as well as for future generations. (NADBank 1998a, 4)

The Mexicali project provides an example of the way BECC approaches public participation:

The first Public Meeting was held . . . with 700 people in attendance. At the meeting it was determined that the majority of the population supports the project. However, a group of local farmers from the Farm El Choropo, though they support the project, raised their concern regarding the location of the Mexicali II treatment plant. The ejido, located between 500 and 1,000 meters away from the plant site, is home to approximately 250 residents. It was apparent there was a lack of information among this group of local farmers, represented at the meeting by almost 20 people. It must be noted that this village is located outside the city limits, approximately 10 km away. (State Public Services Commission of Mexicali 1997, 46)

In the Mexicali project, the public was given an opportunity to reject the completed wastewater treatment plan at two open meetings. This opportunity marks an important difference between the BECC/NADBank and the IBWC. This opportunity for public participation built into the process responds to the pluralist discourse by providing procedures for citizen approval of tax-funded projects. To some extent, offering the public the opportunity to participate also responds to the communitarian call for an open dialogue among citizens on shared concerns.

Some individuals involved with wastewater project planning in Mexico lament the move away from the simplicity of past ways. They have suggested that if the public had not been asked for their comments, the Mexicali II project would probably have been built without any dispute. But these officials nevertheless saw public participation as an unstoppable trend and felt that, for better or worse, the Mexicali project set a precedent for the future in terms of public participation in the process (Int. 87, confidential interview, March 1999). As public participation will be an unavoidable part of their job for the foreseeable future, these officials were interested in doing a good job at encouraging and facilitating it. They also felt, however, that they do not yet know how to create a good forum for public participation, because public participation of this sort is a new process in Mexico (Int. 87, confidential interview, March 1999).

Another innovation instituted by the BECC and NADBank to promote public participation is an electronic list serve (BECCnet). In chapter 5, Levesque analyzed the potential that electronic communication networks hold for the creation of alternative discourses. As in the Y2Y case, the list serve technology has created an important forum for communication, critique, and consensus building. This participatory venue removes the time limitations and agenda-setting processes that often limit the terms of debate. Any interested individual, private-sector consultant, nongovernmental organization representative or government official, or other interested party can subject all of the others on the list to a soliloquy or invitation to debate of any length. The participants set the rules themselves, for the most part, and learn from each other. The majority of the people on the list serve are silent observers. A number of committed and active citizens and government officials routinely post critiques of and suggestions for the BECC and NADBank and the projects as they go through the planning, certification, financing, and implementation processes.

Critique of the BECC and NADBank

The BECC and NADBank were created in response to criticism of the managerialist discourse used by the IBWC in dealing with water management issues in the border area. In justifying their programs, the BECC and NADBank have incorporated some communitarian elements into their predominantly managerialist worldview. As discussed above, the BECC and NADBank also view water as a source of community building.

With regard to the work of BECC and NADBank to build on the potential of water to strengthen the vitality of a community, an observer of the BECCnet described the institutions' contribution to public participation and democracy in the U.S.-Mexico border area as follows:

Anybody who spends any time on the border knows that 16 projects certified in two years, four constructed, two more constructed with non-NADBank money, is a 10-fold increase in what's been going on down there. And, at the same time a process that is empowering citizens, even to just be mad.... It's their business to work it out. I'm glad that they have the opportunity to work it out. (Int. 5, confidential interview, August 1997)

Through the BECCnet, public meetings, web pages, press releases, and distribution of printed material, the BECC and NADBank have yielded some control over setting the terms of debate to local government, nongovernmental organizations, and individual citizens. Still, the BECC-NADBank project cycle does not yet include public participation in the problem definition process.[9] One border resident familiar with the Mexicali project describes the BECC project participation process as overly circumscribed:

Of course when they ask the people if that is what they need, the people say yes. And they say yes because it is the model of development that they know. It is the aspiration that they have been sold. But can we call that participation? ... Have the people really done an evaluation, have they been able to imagine a different future, have they been able to imagine what it is that they really need to participate in a discussion about what would really improve their quality of life? (Int. 84, confidential interview, March 1999)[10]

In the Mexicali project, community involvement occurred only after the planning process was complete. Thus, local citizens did not have control over whether (and how) to frame the problem.

By viewing water as a commodity, the BECC and NADBank provide another potential avenue for citizens to express their preferences. Rather than encouraging learning and communicative exchange toward developing a shared perspective on a course of action, viewing water as a commodity affords a more individualistic approach to decision making. According to the market model, the individual has the option to exit if he or she does not wish to purchase the water/wastewater arrangement offered by the BECC-NADBank system. The value of the individualist approach, in theory, is that individual freedom and voice is maximized, thereby facilitating communication of demand for services and arrangement for their provision.

The economic approach assumes users are willing and able to pay for the cost of a particular water project, with the underlying implication that if they are not able to pay the cost, they should either move to a location where water services are less costly or go without adequate services. If they are able but not willing to pay the cost, then demand is not high enough to sustain functionality of the project, and it should not be built. Accordingly, the approach suggests, scarce investment resources should go to communities with more willingness to pay. Allocation of money in this way is an attempt to increase awareness of environmental constraints on wastewater treatment projects. Critics contend that this approach fails to address certain issues. For example, do those who are unable to pay really have an exit option? What are the ramifications of their "choice" to forgo wastewater treatment plants?

One disadvantage of viewing water as a commodity is that those with less mobility and less income are less able to voice their preferences and more likely to have choices and costs imposed on them. Thus they are likely to be subjected to projects that they do not see as bettering their own lives and, as a result, they are unwilling to contribute to project cost or upkeep (Salmen 1992, 1).

Another problem with seeing water as a commodity is that the noncommensurable aspects of water become less visible. In chapter 3, García-Acevedo chronicled the devastation to the Cocopa people caused by shifts in the meaning of water and the reduced availability of Colorado River water perpetuated by these shifts. Some view the planned cementing of the All American Canal as yet another diversion of Colorado River water from the water-poor Mexicali Valley, as it will prevent infiltration of Colorado River water into the Mexicali-Imperial Valley aquifer used to irrigate crops in Mexicali (Int. 85, confidential interview, March 1999).

The least innovative aspect of the meanings of water articulated by the BECC and NADBank is their reliance on product-type solutions to water problems. One strength of product-type solutions is that they bring expertise to bear on water quality problems. Both BECC and NADBank provide technical assistance to help communities benefit from such expertise.

One problem with constructing water as a product, however, is that it weakens our ability to imagine alternatives to the manufactured present. Because they emphasize a product-oriented type of thinking about water in their organizational legitimation, the BECC and the NADBank tend to support projects designed and constructed by engineering firms rather than household- or factory-level alternatives.[11]

In addition, some groups in Mexicali and the border region see water as an integral part of the ecological and cultural identity of communities. Rather than reshaping or merely utilizing water, these groups believe it is important to learn from the water that helps to form life in the border region. For example, an activist and scholar in the border region, G. Arturo Limon, asks that we "learn from nature ... those old rivers that, calmly go undulating through our Tarahumara mountains, with the patience of one who knows that his final destination is the sea. ... We are part of the earth and she too is part of us" (Limon 1997, 112–113).[12]

Limon views water as a gift of nature rather than an industrial product or an item for purchase. Given the scarcity of water in the environment in which they work, the BECC and NADBank may not be sufficiently sensitive to such alternate perspectives. The fact that the BECC and NADBank are rationalized, in part, by the concept of sustainable development makes such insensitivity somewhat surprising.

Sustainable development is a term whose meaning is contested. Scholars point out the inherent contradiction of sustaining growth in material consumption on a finite planet (Daly 1996). The U.S. government sees the path toward sustainable development in terms of incremental changes guided by a new focus in the way that interactions among economic, environmental, and social processes affect both current results and long-term endowments and liabilities (SDI Group 1998). The World Commission on Environment and Development (1987) saw it as an urgent break from "business as usual," which will require difficult choices between economic growth and environmental preservation in industrialized as well as in developing nations (SDI Group 1998). Environmentalists hesitate to use the phrase at all because they believe multinational businesses are using the concept of sustainable development to obfuscate anti-environmental activities (AtKisson 1999, 134).

Although sustainable development is characterized by ambiguity and conflicting definitions, many definitions of sustainable development emphasize the idea of striving to emulate and protect self-renewing ecosystems. They also stress reliance on locally available resources rather than importation of resources from elsewhere, along with the value of in situ ecosystemic health. The EcoVillage at Ithaca, New York, is an example of this self-sufficiency view of sustainable development.[13]

Other commentators on sustainable development emphasize ethical or spiritual integrity more than environmental concerns. For example, Ken Conca (1996) suggests ways that peace and justice can form pillars around which sustainable societies can be built. On a more spiritual note, Herman (1988) concludes that "We cannot do anything to other people and to nature without simultaneously doing it to ourselves.... Towards improving the state of the world, then, it is up to individuals to take the initiative and do their part.... Self-realization on the individual level is necessary for such a transformation on the global level" (278–280).

Embedded in the concept of sustainable development is a struggle between the maintenance of current ecological, social, and economic conditions and their sustained improvement, and whether and where trade-offs should be made to advance one at the expense of the others. Herman Daly and John Cobb (1994) suggest that at a minimum, "the running down of natural capital would have to be offset by the accumulation of an equivalent amount of humanly created capital" (72). Preferably, a "strong sustainability" should be adopted. Strong sustainability "would require maintaining both humanly created and natural capital intact separately, on the assumption that they are complements rather than substitutes in most production functions" (72). Pirages (1996) offers another spin on this ambiguity: "Building and living in more sustainable societies ... need not require general ecological penance such as force-feeding humanity vegetarian diets and creating a fashion industry based upon sackcloth.... Real intellectual excitement can be generated over devising ways to do more with less" (8).

These commentators on sustainable development imply that recognizing water as a gift of nature can help communities attain a high quality of life through the consumption of fewer material and natural resources. Sustainability requires innovative solutions to ecological constraints and cultural priorities that are not readily apparent when water is viewed primarily in modernist terms. Incorporating other views of water can augment efforts to achieve sustainable development.

Moving toward a more sustainable lifestyle is likely to be quite difficult where existing quality of life and levels of consumption vary as widely as they do in the border region. Out of necessity, many low-income families on the Mexican side of the border have become quite expert at low-cost water conservation and reuse techniques. Their experiences, both positive and negative, should inform efforts to effect more sustainable and equitable water usage on both sides of the border.

The current efforts of the BECC and NADBank constitute a partial response to the water-related problems along the U.S.-Mexico border. Whether the BECC and NADBank will be able to incorporate more elements of the communitarian discourse into their operationalization of sustainable development is questionable. The communitarian discourse is largely absent from decision-making structures of the two organizations. The BECC may become more open to meanings of water linked to the communitarian view if its governing structures and constituents see value (social, political, and financial) associated with such meanings. The forums for public participation associated with the BECC can facilitate communication of this value if efforts are made to bridge the managerialist, pluralist, and communitarian discourses. It may be more difficult for the NADBank to make this sort of change, as it does not have a public advisory committee or similar forum to facilitate the difficult process of interdiscursive dialogue.

Conclusion

The range of concerns and needs arising around international bodies of water is a product of environmental, economic, and social history. As such it is as unique as water bodies themselves. Drawing upon Williams and Matheny (1995), this chapter has illustrated the contribution that discourse analysis can make to the study of the meanings of water in international contexts. It is the social construction of a problem that defines the action that appears most appealing to the actors involved, whether they select individual gain, due process, or equity as their decision criteria.

The chapter has traced the way in which the dominant managerial discourse was challenged and replaced by a somewhat modified yet still largely managerial discourse, concluding that, despite shortcomings, a more open discursive arena now exists as a result of the construction of water as a source of community building by the BECC and NADBank. At the same time, the BECC and NADBank have been shown to view water as a product and commodity. If the goal is to accomplish ecological, economic, and social sustainability in the U.S.-Mexico border region, then such views of water may not be sufficiently sensitive to alternative premodern and postmodern conceptions described in chapter 2.

Notes

1. Specifically, I draw upon the following sources for my analysis: (a) The Agreement between the Government of the United States of America (U.S.) and the Government of the United Mexican States (Mexico) Concerning the Establishment of a Border Environment Cooperation Commission (BECC) and a North American Development Bank (NADBank), November 1993 (entry into force Jan 1 1994); (b) The *BECC Project Certification Criteria* (BECC 1996); (c) 1998 NADBank promotional/programmatic materials, general information, loan and guarantee program description; (d) confidential interviews with BECC and NADBank staff, interested World Bank and Inter-American Bank staff, interested federal officials in the United States and Mexico, interested consultants in the United States, and environmental nongovernmental organization activists in the United States and Mexico (seventy-eight interviews total); and (e) a limited survey of interested border residents in the U.S. and Mexico (twelve respondents).

2. The treaty governs utilization of waters of the Colorado and Tijuana Rivers and of the Rio Grande. It was signed in Washington on February 3, 1944.

3. Wildavsky (1979) argues that policymakers define problems to fit available and politically acceptable policy actions by government. Thus, problems do not precede solutions, but rather solutions precede problems.

4. Agreement 1993 (*supra*, note 1(a)), Chapter I, Border Environment Cooperation Commission, Article I, Purpose and Functions, Section 1. Purpose.

5. Agreement 1993 (*supra*, note 1(a)), immediately after entry into force of the North American Free Trade Agreement.

6. The BECC was not involved in early phases of these projects and hence was unable to encourage or facilitate completion of the high-sustainability approval process, but it intends to be so involved in the future (Int. 49, confidential interview, January 1998).

7. According to an interviewee knowledgeable on this matter, a statement on September 18, 1998, by the principal engineers of the IBWC/CILA specified this intention.

8. Agreement 1993 (*supra*, note 1(a)).

9. For an analysis of the importance of listening to beneficiaries' views in the problem definition phase, see Salmen 1992, 1987, and Robb 1998.

10. Author's translation of the following: "Por supuesto que cuando ellos le preguntan a la gente si eso es lo que necesitan, la gente dice que si. Y dice que si porque es el model de desarrollo que conocen. Es la aspiracion que les han vendido. Pero a eso le podemos llamar participacion?... [R]ealmente la gente ha hecho una evaluacion, ha podido imaginar un futuro diferente, ha podido imaginar que es lo que realmente necesita participar de una discusion acerca de que es lo que realmente mejoraria su cualidad de vida?"

11. Pombo (1999) shows that there is an unmet demand for household-level wastewater treatment alternatives in Tijuana, a city within the service area of the BECC and NADBank. Because these organizations view wastewater as a product, they have not adopted such intermediary approaches.

12. Author's translation of the following: "Aprendamos de la naturaleza ... esos viejos rios que calmados van ondulando nuestra sierra Tarahumara, con la paciencia de quien sabe que su destino final es el mar.... Somos parte de la tierra y asimismo ella es parte de nosotros."

13. EcoVillage maintains a Web page at ⟨http://www.cfe.cornell.edu/ecovillage/⟩. See Lachman 1997.

References

AtKisson, Alan. 1999. *Believing Cassandra: An Optimist Looks at a Pessimist's World*. White River Junction, VT: Chelsea Green Publishing.

Audley, John J. 1997. *Green Politics and Global Trade: NAFTA and the Future of Environmental Politics*. Washington, D.C.: Georgetown University Press.

Barber, Benjamin R. 1998. *A Place for Us: How to Make Society Civil and Democracy Strong*. New York: Hill and Wang.

BECC (Border Environment Cooperation Commission). 1996. *BECC Project Certification Criteria*. Ciudad, Juarez, Mexico: Border Environment Cooperation Commission.

Conca, Ken. 1996. "Peace, Justice, and Sustainability." In Dennis Pirages (ed.), *Building Sustainable Societies: A Blueprint for a Post-Industrial World*. New York: M. E. Sharpe, pp. 17–31.

Daly, Herman E. 1996. "Physical Growth versus Technological Development: Odds in the Sustainability Race." In Dennis Pirages (ed.) *Building Sustainable Societies: A Blueprint for a Post-Industrial World*. New York: M. E. Sharpe, pp. 167–173.

Daly, Herman E., and John B. Cobb, Jr. 1994. *For the Common Good: Redirecting the Economy toward Community, the Environment, and a Sustainable Future*. 2d ed. Boston: Beacon Press.

EPA (U.S. Environmental Protection Agency). 1996. *U.S./Mexico Border XXI Program. Summary of Domestic Meetings*. June.

EPA-SEDUE (U.S. Environmental Protection Agency and Mexican Secretaria de Desarrollo Urbano y Ecologia) 1991. *Integrated Environmental Plan for the Mexico-U.S. Border Area (First Stage, 1992–1994). Working Draft*. August 1.

Etzioni, Amitai. 1983. *An Immodest Agenda*. New York: McGraw-Hill.

Fischer, Frank. 1990. *Technocracy and the Politics of Expertise*. Newbury Park, Calif.: Sage.

Hendee, W. R. 1990. "A Permanent United States–Mexico Border Environmental Health Commission." *Journal of the American Medical Association* 263, no. 24 (June 27): 3319–3321.

Herman, R. Douglas K. 1988. "Mind and Earth: A Geography of Being." Unpublished master's thesis, University of Hawaii, Manoa.

IBWC (International Boundary and Water Commission). 1973. Minute 242: "Permanent and Definitive Solution to the International Problem of the Salinity of the Colorado River." International Boundary and Water Commission, El Paso, Texas.

Ingram, Helen, and David R. White. 1993. "International Boundary and Water Commission: An Institutional Mismatch for Resolving Transboundary Water Problems." *Natural Resources Journal* 33 (Winter): 153–175.

Kingdon, John W. 1984. *Agendas, Alternatives, and Public Policies*. Boston: Little, Brown.

Lachman, Beth E. 1997. *Linking Sustainable Community Activities to Pollution Prevention: A Sourcebook*. Santa Monica, Calif.: RAND.

Limon, G. Arturo. 1997. *Hacia una Cultura Ecologica en Chihuahua* [Toward a Cultural Ecology in Chihuahua]. Chihuahua, Mexico: Talleres Graficos del Gobierno del Estado de Chihuahua.

Lindblom, Charles E. 1979. "Still Muddling, not Yet Through." *Public Administration Review* 39 (November/December): 79–88.

Mumme, Stephen. 1993. "Innovation and Reform in Transboundary Resource Management: A Critical Look at the International Boundary and Water Commission, United States and Mexico." *Natural Resources Journal* 33 (Spring): 93–120.

NADBank (North American Development Bank). 1998a. *General information brochure.* San Antonio, Texas.

NADBank (North American Development Bank). 1998b. *Loan Guarantee Program.* San Antonio, Texas.

Pirages, Dennis. 1996. "Introduction." In Dennis Pirages (ed.), *Building Sustainable Societies: A Blueprint for a Post-Industrial World.* New York: M. E. Sharpe, pp. 3–13.

Pombo, O. Alberto. 1999. "Environmental Analysis and Socio-Ecological Consequences of Water Use and Sanitation Practices in Marginal Areas of Tijuana: A Demand-Side Perspective." Paper presented at the Center for U.S.-Mexican Studies Conference, "Economic Integration and the Environment: Promoting Sustainable Development along the U.S.-Mexican Border," February 5. University of California, San Diego, La Jolla, California.

Robb, Caroline M. 1998. *Can the Poor Influence Policy? Participatory Poverty Assessments and the Challenge of Inclusion.* Washington, D.C.: World Bank.

Salmen, Lawrence. 1987. *Listen to the People: Participant-Observer Evaluation of Development Projects.* New York: Oxford University Press for the World Bank.

Salmen, Lawrence. 1992. *Reducing Poverty: An Institutional Perspective.* Washington, D.C.: World Bank.

Sanchez, Roberto. 1993. "Public Participation and the IBWC: Challenges and Options." *Natural Resources Journal* 33 (Spring): 283–298.

Schneider, Anne Larason, and Helen Ingram. 1997. *Policy Design for Democracy.* Lawrence: University Press of Kansas.

SDI Group. 1998. *Sustainable Development in the United States: An Experimental Set of Indicators.* Washington, D.C.: U.S. Interagency Working Group on Sustainable Development.

State Public Services Commission of Mexicali. 1997. "BECC Step II Project Format for the 'Sanitation Program of the City of Mexicali.' " October.

TCPS (Texas Center for Policy Studies). 1992. "NAFTA and the U.S./Mexico Border Environment: Options for Congressional Action." Texas Center for Policy Studies, Austin. September.

Utton, Albert E. 1984. "Overview." In Cesar Sepulveda and Albert E. Utton (eds.), *The U.S.-Mexico Border Region: Anticipating Resource Needs and Issues to the Year 2000*. El Paso: University of Texas.

Wildavsky, Aaron. 1979. "Analysis as Art." In *Speaking Truth to Power: The Art and Craft of Policy Analysis*. Boston: Little, Brown, pp. 1–19.

Williams, Bruce A., and Albert R. Matheny. 1995. *Democracy, Dialogue, and Environmental Disputes: The Contested Languages of Social Regulation*. New Haven, Conn.: Yale University Press.

8

A Hydroelectric Power Complex on Both Sides of a War: Potential Weapon or Peace Incentive?

Paula Garb and John M. Whiteley

In the early 1990s, when the former Soviet Union broke up into fifteen independent states, boundaries that people never realized existed inside these former republics suddenly became flash points for tensions and war. Armed conflict over disputed borders and jurisdictions of autonomous ethnic regions inside the post-Soviet republics explosively divided the newly independent states of Russia: Moldova (in the southwest Soviet Union) and Azerbaijan and Georgia (both in the southern Soviet Caucasus). One such ethnopolitical war was fought between Georgians and Abkhaz over Abkhazia's bid for independence from Georgia after the breakup of the Soviet Union. On August 14, 1992, Georgian troops entered Abkhaz territory to take control of its Soviet-era autonomous republic. Fourteen months later Abkhaz forces, with covert assistance from Russia, pushed Georgian forces back across the Inguri River (see figure 8.1). Since that time, peace talks in the region have been deadlocked. Abkhazia has not been recognized by any government but enjoys de facto independence from Georgia.

This chapter focuses on developments involving the Inguri River, a natural water boundary that, with the help of peacekeeping forces from the Commonwealth of Independent States (CIS) and United Nations, physically separates Georgian and Abkhaz forces. It examines how, in the midst of these conflicts, the Abkhaz and Georgians jointly manage the Inguri River hydroelectric power complex, consisting of a dam and five hydroelectric power plants. Components of the power complex are divided between the territories of the two ethnic groups, with the dam on the Georgian side of the river and the power plants on the Abkhaz side. In the Soviet Union, the Inguri was the divide between Georgia and its

Figure 8.1
Abkhazia

autonomous republic but, in the Soviet system, this internal border had little meaning. These facilities, contrary to the vision of the Soviet politicians and engineers who built them in the 1970s, now are related integrally to the deadly struggle between Georgia and Abkhazia.

Both sides regard the water of the Inguri River as a resource vital for the production of an essential commodity, electricity. Water in this case is a security issue, because its use in the production of electricity at the Inguri complex is fundamental to national survival and to the building of the nation-state. As chapters 1 and 2 point out, the many benefits provided by water resources, including boundary marking, supplies for human consumption, navigation, flood control, and irrigation, as well as the many economic benefits associated with them, have long been basic to the existence and survival of the nation-state. Thus, water has historically been associated with nation-state politics and national security. These are the issues of the modern world, not of the postmodern context described in most of the other chapters of this book.

Water can be either a force for fanning the flames of conflict or for dampening them (Lowi 1995). Although the Inguri facilities do have value as a weapon, any unilateral action to prevent electrical production or to destroy the complex would renew the conflict and ignite full-scale war. This makes the Inguri complex a transboundary parallel to the cold war doctrine of mutually assured destruction.

In this case, cooperation exists in the midst of conflict. In the sections that follow, we intend to explain this contradiction. We will show that modern approaches rooted in rational behavior continue to provide considerable insight here as they have in other cases (chapter 9). At the same time, the Inguri case defies easy classification. The essentialistic attachment ethnic groups have to land and water in this region is premodern. However, this case also demonstrates some rudimentary transboundary ties, which are characteristic of postmodern contexts (chapters 4 and 5).

Understanding how cooperation can take place under such unlikely circumstances may provide insights that promote movement toward the resolution of other highly charged border conflicts. The Inguri River power complex provides a model of coerced, but effective, cooperation. The potential for cooperation is especially compelling when addressed in the context of deadly armed conflict and its aftermath, the constant threat of resumed conflict, and an ongoing struggle to curtail random terror, assassination attempts, and human rights abuses. This case of a contested transboundary resource, regarded as both a product and a security issue, serves as a useful prism through which to refract and identify the conditions necessary for cooperative transboundary water management.

The Inguri facilities are an integral and absolutely essential element of the state building activities of both conflicting parties. This case is unlike that of Lake Constance (chapter 4), where one of the first and most successful examples of international environmental regimes was formed. At a time when international networks are surging forward, the Inguri case tells us that not all nation-states are fading into the background. The Black Sea Environmental Programme (BSEP) (chapter 9) also shows that states are not necessarily disappearing as regional arrangements like the BSEP emerge. The BSEP agreements, which established that organization, also prevent it from working with nonstate actors, such as the unrecognized territory of Abkhazia. In many transboundary water issues, states continue to be very important.

To make sense of this case, with all its complexities and contradictions, a contextual and nuanced analysis is essential. The ethnographic and historically informed case study presented in this chapter reveals the culture-specific and historical factors involved in the conflict in this region and the roots of local identities and relationships that have an impact on the bilateral management of this transboundary water resource and the related peace process. The information and analysis are based on in-depth interviews with officials, managers, and employees of the complex, as well as with outside observers in the two communities. Maintaining the confidentiality of the informants is particularly necessary because of the sensitivities involved in providing information about this high-security area. Another source of information is reviews of newspaper articles in the region.

The chapter is divided into three parts. We begin by providing the social and historical context for our discussion by tracing the history of the conflict and the cultures of the border communities. In the chapter's second section, we describe the Inguri River, the dam and the associated electric power generation, and the cooperation that exists between the Georgians and Abkhaz in managing the complex. The final section explores multiple explanations for this cooperation.

Cultural and Ethnic History of the Inguri Region

Georgians and Abkhaz both identify with and assign important meaning to the region around their joint border and its rivers and streams. Both believe that this is their historical homeland. However, until the 1992–1993 war, these people lived in peace as neighbors and relatives, as they had for most of their long history, even though they are two distinctly different ethnic groups with different languages and cultures. During the Soviet era, Stalinist policies pitted the two groups against one another, fomenting deep-seated hostilities. Tensions rose so high that, by the end of the Soviet era in 1991, the Abkhaz feared that the Georgians were intent on invading what they consider to be long-established Abkhaz territory and conducting cultural and physical genocide against them. The Georgians held conflicting beliefs. They believed the Abkhaz were taking away an important portion of their ancient homeland. The result has been an identity-based war, which is usually a zero-sum game. Even now, after armed conflict has ceased, the two sides still do not agree about these critical issues of territory, history, and identity, and so they remain at a standoff.

The war, which was fought entirely on Abkhaz territory, left 4,000 Abkhaz and 10,000 Georgians dead and resulted in at least 200,000 refugees. Almost all of these refugees are ethnic Georgians formerly of Abkhazia, now living over the border in Georgia proper. The fighting destroyed most Abkhaz industrial facilities, along with other important institutions such as the university, the drama theater, the research institute on the Abkhaz language, literature, and culture, the central state archives, the physics and technology institute, and the parliament building. During the assault there was much destruction of Abkhazia's homes and physical infrastructure (e.g., bridges, sewage treatment facilities, the railway). An estimated 30,000 land mines still cover the Abkhaz countryside, including large areas around the Inguri complex (Paddock 1999). It is estimated that it will take several decades to clear all the mines, since no maps were made of their locations. The United Nations has calculated that it will cost a minimum of $187 million to rebuild the most essential aspects of Abkhazia's infrastructure (United Nations Mission on Needs Assessment in Abkhazia, Georgia 1998).

Although the dam and power plants have suffered neglect during this conflict because only minimal resources were available for regular maintenance and repairs, neither the Georgian nor the Abkhaz military staged any of their battles in the vicinity of the facilities. This can be attributed first to the fact that the Gali region, where the plants are located, was not a military arena until the very last days of the war in September 1993 when Abkhaz troops, facing little resistance, were pushing out the Georgian forces. Second, by all accounts on both sides, both sides recognized that the Inguri facilities were far too important to destroy deliberately.

To date, the Georgian-Abkhaz conflict has not been resolved. CIS military forces guard the border at the Inguri River while the parties in conflict negotiate a settlement, with assistance from the United Nations, Russia, and Western powers. On Abkhazia's border with Russia at the Psou River, Russian border officials enforce a blockade that severely restricts the movement of people and goods. They also prevent ships from entering Abkhazia's Black Sea port of Sukhumi. Despite efforts to achieve peace, terrorist acts and skirmishes that result in more refugees and more suffering erupt periodically in the southern part of the Gali region (the power plants are in the north), where Megrels are the dominant ethnic group.

Megrels are one of several Georgian subethnic groups that converse not only in Georgian, but also in their own related language. Most Megrels have a dual identity as both Georgians and Megrels. (In this chapter we refer to these people as Megrels to distinguish them from other Georgians and to identify them as they are known in the border communities.) Megrels are indigenous to western Georgia and have always lived as neighbors with the Abkhaz. Abkhaz and their language are not related to either Megrelian or Georgian; neither group of speakers can understand the other's language. Abkhaz identify themselves as an ethnic group separate from both Megrels and Georgians. Because they have lived so close to one another for such a long time, Megrels and Abkhaz have similar cultural practices and mentality. However, many distinct differences between them create gaps in cross-cultural understanding.

Populations have shifted over the centuries in this border area, with each of these two unrelated ethnic groups having experienced considerable assimilation into the other. In antiquity, Gali was most likely populated by early Abkhaz tribes, judging by the names of the smaller geographical sites (small rivers, streams, etc.). This suggests that Abkhaz tribes were in the area long before there was any recorded historical information. Small geographical sites, especially rivers, tend not to be renamed when a population changes and are therefore a good indication of the culture of the earliest populations. Even the historical names of larger geographical sites indicate the early presence of Abkhaz tribes. For instance, the ancient name of the Rioni River, the largest river in western Georgia, was Fasis, which is of Abkhaz origin. Absar was the earlier name of the Chorokha River in western Georgia, and the Supsa has always been the name of a river in that same area. Both of these names are also of Abkhaz origin. These are indications of the presence of Abkhaz tribes in western Georgia as early as the first century A.D. The Abkhaz name for the Inguri River is the Egry, which is pronounced almost the same as the Megrel name for the river, evidence that both groups have considered this border area their homeland for a very long time. These linguistic formulations show how deeply geography and water systems are imbedded in culture.

The line that separates Abkhaz and Megrel territories has been unstable over the centuries. Historically, water has always played an important boundary role in this region as well as in others (chapter 10). At the beginning of the seventeenth century the ethnic and political border was around the Kodor River (which is far inside Abkhazia in the contemporary Ochamchira region). In the second half of the seventeenth century, Abkhazia conquered the area up to the Inguri River. Gali (called Samurzakan since the Abkhaz began to dominate) has had a mixed population of Abkhaz and Megrels for more than a century. The upper class in Gali was composed entirely of Abkhaz, whereas the peasants were of both ethnic backgrounds. Language assimilation and usage in Gali prevailed according to the areas of geographical dominance by each ethnic group: Abkhaz in the north and Megrels in the south.

In the second half of the nineteenth century, the influence of the Abkhaz language weakened as Abkhaz were increasingly assimilated by the Megrel culture subsequent to Abkhaz emigration to the central region and exile in Turkey. In addition, the Megrel conversational language gained dominance as peasants from the left bank of the Inguri, seeking greater freedoms, migrated to the right bank, where weaker practices of serfdom prevailed. By the second half of the nineteenth century, Megrel was the dominant language in the area. Although the vast majority of people throughout the Soviet period identified themselves as Georgians and/or Megrels, many claimed very mixed heritage due to intermarriage. In September 1993, when Abkhaz forces advanced as far as the Inguri River, the predominantly Megrel population in the river region became victims of war rage. To escape the wrath of the Abkhaz for Megrel crimes committed in central Abkhazia, most Megrel populations and other Georgians fled from the Inguri region and the rest of Abkhazia.

Once the fighting ceased, Megrels began returning to their homes in Gali, which is administered and policed by Abkhaz authorities, while CIS troops stand guard at the border and UN military observers patrol the region. A percentage of the former population of Gali, together with Megrels from the adjacent Zugdidi region and other Georgians, have been engaged in terrorist acts in an attempt to secure Gali, if not the rest of Abkhazia, and to bring it back into Georgia. In May 1998, a major attempt by terrorists to retake Gali failed. By this time thousands of the earlier Georgian refugees had returned to their homes and revitalized their farming. Once again the same people had to flee, this time leaving behind villages to be burnt by Abkhaz soldiers. Tensions are still quite high at the time of this writing, and the fate of Gali remains unresolved.

The situation in the area is further complicated by the culture of blood revenge, which the Abkhaz still practice widely. According to tradition, the culture of blood revenge compels family members of someone who has been killed to take revenge against the relatives of the offenders. In case of war, when it is not always possible to determine the murderer, it is acceptable to kill any member of the murderer's ethnic group. In such a culture, responsibility for murder is not individual but collective. Megrels also practice blood revenge, but it is not so widespread today, and the target of Megrel revenge is usually the perpetrator of the crime, not family members or a whole clan. Not all members of any of the three ethnic groups involved condone and practice the blood revenge tradition, but enough of them do to make postwar healing far more complex than in cultures without traditions of patterned revenge killing. Within this historical and cultural context, the Inguri hydroelectric power complex was built in the late 1970s and early 1980s.

A History of the Inguri Complex and Current Cooperation

At the time of construction of the power complex, these ethnic and political considerations did not figure in the calculations of the Soviet central government, the Georgian and Abkhaz authorities, or the engineers who designed the facility. Water was disembedded from historical and cultural considerations and became simply a means to a product. Just as in the case of the Mexicali and Imperial Valleys presented in chapter 3, where water wealth produced agricultural crops, water here generated hydroelectric wealth. The Inguri complex was built to be an integral part of the national Soviet power grid and was intended to provide supplemental electricity during periods of peak demand. Electric power produced at the Inguri Dam, as at all such industrial complexes, was centrally managed in Moscow (Carlisle 1996).

The Inguri complex, the largest among the many Soviet hydropower facilities was built on the Inguri River, which is 221 kilometers long and originates on the southern slope of the Caucasus range. The river flows downstream from the mountains into narrow and deep canyons that provide perfect dam sites. Alongside the town of Dzhvari (in the Megrelia region of Georgia), where the river flows over flat country, the Inguri River branches off and runs into the Black Sea in the Gali region. The Inguri combines a significant volume of water flow with a steep slope of descent, both of which are critical to hydropower generating capacity. The total elevation decline over the river's course is 2,600 meters.

The Inguri complex (see figure 8.2) is situated on both sides of the boundary delineated by the Inguri River and consists of the world's largest arch dam and reservoir (on Georgian territory) and five interconnected power plants (on Abkhaz territory), including the main diversion dam unit, a dam-type plant, and three similar river channel plants that discharge into the Black Sea (on Abkhaz territory). The Inguri plants were built with an electricity production capacity of 1.3 million kilowatt/hours (Golubchikov 1995). Today they produce only from 200,000 to 600,000 kilowatt/hours because the facilities are in such bad repair. From the higher elevation of the main dam (in Georgia proper), water is diverted through a large channel extending for fifteen kilometers into Abkhaz territory. Twelve of the fifteen kilometers are on the Georgian side of the Inguri. The falling water generates electricity on its way from the mountains of Georgia through the mountains of Abkhazia to the Black Sea.

In the Soviet Union the plants served as a key source of hydroelectric power for the Southern Caucasus power grid. They contributed peaking power to meet demand as far east as Kazakhstan. Soviet leader Nikita Khrushchev originally wanted the facility to be built on the Bzyb River, which flows from majestic mountains into the Black Sea through the region of Khrushchev's beloved Abkhaz vacation spa in Pitsunda. Engineers, however, persuaded him against this idea because of the

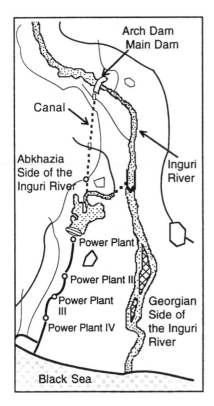

Figure 8.2
The Inguri complex

disastrous environmental and economic consequences of beach erosion in this prime resort area. Instead, the Inguri River was chosen because it flowed into a section of the coast that was not conducive to resort development, so beach erosion was not regarded as a prime concern.

When the project finally went on line in the late 1960s and early 1970s, it gave a major boost both to the state sector of the economy and to the private sector (the illegal, shadow economy) in the Zugdidi (Georgia proper) and Gali (Abkhazia) regions. It employed large numbers of people in relatively high-paying jobs. Migrants from central Georgia moved to the area in large numbers, requiring housing and infrastructure development. The inflated budgets of all the new construction sites also provided construction materials and money that were illegally siphoned off from the Inguri project. Thus although the complex was a major facility in the national Soviet system, it also played a key role in the development of the local economy of the Gali region.

At the beginning of the armed conflict neither side was focused on the Inguri facility as an asset. The fighting began over a complex set of identity, language, and history issues overlaid with disputes over economic resources, such as those provided by sea resorts and lucrative cash crops. The Inguri River and its electricity did not become part of the calculation of war until after the actual fighting had ended, when the forces of both sides found themselves facing each other from separate sides of the river, each holding on to one part of this huge complex, which was totally useless without both sections working together.

Looking closely at how the joint management of the complex operates in practice provides insights into the factors that have facilitated multiple levels of cooperation, that is, between the Georgian government and the de facto Abkhaz government, between the managers of the respective energy agencies on both sides of the conflict, and among the plant employees. Inextricably linked to other unsettled issues in the peace process, the ownership of the complex is still undecided. Primary among these issues is Abkhazia's political status. Discussions on property rights have thus far focused on joint ownership, seeking to ensure continued power generation regardless of the outcome of the final decision on Abkhazia's status as either a recognized independent state or a political entity within Georgia. The two sides, however, fundamentally disagree over the criteria to be used in determining the value of each structure and the relationship of ownership value to the income derived from generated electricity.

The Abkhaz insist that ownership value should be determined by the original cost of building the physical structures. In this case, they assert that 40% of the ownership value should be on the Abkhaz side and 60% on the Georgian side. Thus the Abkhaz are persuaded that generated electricity should be distributed accordingly on a 40/60 split. The Georgians, on the other hand, argue that the criterion for the complex's value should be the percentage of the total electricity generated by the complex that each side consumes. Therefore, since Georgia estimates it can use 80% of total generation, Georgian representatives assert that the two sides should divide ownership value 20/80. The Abkhaz counter-argument is that Abkhazia's needs for electricity will increase with economic development, and that if it has more electricity than it needs, all the better: It can sell the surplus to Russia, Georgia, or someone else. At this time, this issue remains unresolved.

Meanwhile, the de facto arrangement is one of joint management, leaving aside the issue of final ownership. A top-down management system is in place; the presidents of both sides, Vladislav Ardzinba (Abkhazia) and Eduard Shevardnadze (Georgia), make or approve all final decisions. High- and low-level staff are mostly ethnic Georgians or Megrels, most of whom worked at the site before the war. There is no formal structure of joint management. The closest entity to a joint structure is the Georgian-Abkhaz Coordinating Committee on Practical Issues, established in November 1997 under a UN-facilitated agreement reached at the Geneva peace talks. The Georgian side is represented on this committee by the Georgian state chancellor (second to the president), the minister of fuel and energy, the minister of communications and mail, the head of the Department of Communications, and the president of the Georgian International Oil Corporation. On the Abkhaz side the representatives are the prime minister, the first vice prime minister and minister of the economy, the personal representative of the Abkhaz president, the general director of the Abkhaz State Building Company, the general director of the Chernomorgenergo State Stockholding Company, and the general director of the State Communications Company.

The Georgian Ministry of Fuel and Energy and the Abkhaz Chernomorgenergo State Stockholding Company jointly appoint the director of the Inguri complex. These two entities approve all important decisions on a daily basis, meeting by telephone several times a day. Even though the director is an ethnic Georgian (Megrel), the Abkhaz are adamant about his staying on the job even if the Georgian government were to want to replace him. The director of the complex and the head of Abkhazia's Chernomorgenergo met and started working together forty years ago in the Georgian power industry. Their long-time relationship and complete personal trust is essential to the smooth operation of the complex in these difficult circumstances. Although such a relationship of trust would be of tremendous value anywhere in the world, it is essential in the cultures of the Caucasus.

In addition to daily telephone conversations, the top managers (excluding President Ardzinba of Abkhazia and President Shevardnadze of Georgia) from both sides travel to the complex on an average of twice a month, where they meet face to face with the on-site director and staff to discuss matters that cannot be handled satisfactorily by telephone. The Russian commander of the CIS forces is always present at these meetings. Since September 1997, these managers have also been coming together for UN-facilitated coordinating committee meetings, which also involve the commander of the CIS forces. The de facto joint management arrangement was further reinforced when the European Bank of Reconstruction and Development required the sides to write a joint memorandum on cooperation before it would consider financing reconstruction projects involving the facility.

This particular arrangement of joint management also functions as a means of relationship building, which is critical to a peace process (Lederach 1997). The subject matter of the power complex has forced the two leaders (Ardzinba and Shevardnadze) to talk to each other, by telephone and in person, many more times than they would have otherwise. These constructive conversations and the resulting joint decisions have reinforced the possibility for positive interaction that can possibly facilitate the peace process and influence the political negotiations. Managers from both sides have stated that they believe the collaboration between Shevardnadze and Ardzinba over the Inguri complex has had a positive impact on peace building.

Explanations for the Success of Cooperation

The primary explanation for the success of cooperation at the Inguri complex is a modern, rational one: The complex is simply too vital for either side to lose. Both sides fear the consequences of a lack of cooperation, that is, that either the electricity will be turned off or water will not flow to the power plants. Electricity from the Inguri complex continues to benefit both the Abkhaz and the Georgian populations. Authorities on both sides have agreed to the terms of financing and joint management of the complex, even though they cannot agree on the terms of a peace settlement. These leaders are, by their actions on this issue, modern rationalists. They act to preserve their own rational, economic interests and those of their people. Therefore, they are motivated more by a modern, utilitarian logic about water as a security issue for building the nation-state.

For the Abkhaz, the Inguri facilities are key to their efforts to construct an independent nation, to rebuild basic infrastructure ravaged by war, and to maintain economic and social well-being. This key role of the Inguri complex is due first to the fact that it is their only source of electricity, serving all their needs. Second, as long as the Abkhaz have the power plants under their control and can threaten to withhold electricity supplies, they are better able to maintain de facto independence. Such leverage is especially important because Abkhazia, unlike Georgia, is isolated from the international community. Abkhazia is not currently recognized by any other nation in the world. Without such recognition, the ability to hold and control a part of the facility is critical to the empowerment of the weaker downstream territory (Williams 1996). For the Georgians, the facility is also essential, since it provides them with most of their electricity for state building. If the Georgians had full control of the entire complex, they might have greater influence in persuading the Abkhaz to remain part of the Georgian state.

Both sides have used the site as a weapon against the other. Thus coercion is a significant ingredient in this delicate equilibrium. For instance, in April 1997, Abkhaz authorities cut off electricity to Georgia in response to the Georgians having turned off the long-distance phone service in Abkhazia. This mutual coercion resulted in a resumption of talks, which subsequently led to the restoration of electricity to western Georgia and to restoration of limited long-distance phone service in Abkhazia. Georgian terrorists have continued, however, to attack electricity lines on Abkhaz territory to interrupt service.

The river and the facility clearly have enormous military and strategic importance for both sides: Electricity is essential to rebuilding the devastated economies of the total region by making state building projects possible in both Georgia and Abkhazia. Therefore, it is reasonable to expect the Inguri complex to be part of the discourse on military and security issues. Its absence from such discourse, however, is striking. The media in both territories regularly discuss disagreements about who holds legitimate historical and cultural claims over the Gali region, about how to solve the problem of the refugees from the Gali region, and about other political and economic issues unrelated to the potential of miltary action at the Inguri complex. If the Abkhaz and Georgian media mention the Inguri site, it is only in terms of how the complex is a positive model of joint cooperation.

In fact, we found no public discussion that communicates an identification of the Inguri complex as having military and strategic significance. Engineers involved in the management and operation of the complex on each side appear sincerely not to see the issue in the context of contemporary politics of the regional conflict. One exception was an Abkhaz official who told us that the Inguri power plants are worth more than all of Abkhazia's assets and losing them would mean losing everything. Few other people interviewed on either side, however, thought the complex was among the main issues at stake.

We hypothesize that the power plants are so pivotal and essential to ongoing affairs that they stand far apart from, and far above, other issues. Few people even think to discuss them in relation to potential conflict, even in private. The exception may be a handful of top-ranking politicians on either side who are willing to discuss the issue. At the end of 1998, the matter was so sensitive that the Georgians did not even want to appear to be encroaching militarily on territory close to the complex in the northern part of the Gali region. All the battles in the Gali region up to that time had been in the southern area, far from the facilities in the north. Perhaps the Georgians were worried that if they even appeared to be preparing to try to seize the plants, the Abkhaz would simply blow up the facilities and they would lose everything. It seems reasonable to speculate that periodic fighting in the Gali region is largely due to both sides' desire to have control over the Inguri power plants. If the Abkhaz hold on to the plants they maintain equilibrium. If they lose them, probably the Georgians triumph completely, and the Abkhaz will have no bargaining power whatsoever.

The river and the facilities continue to be important as long as no other major source of electricity exists for either side, which seems certain to remain the case. At this level of analysis, the parties are quite rational. It is not rational to take action that threatens oneself as much as the enemy. Both sides equally condemn terrorist acts against the facility or permanent subversion of the distribution of electricity. Neither side has even appeared to be planning, let alone carrying out, efforts to destroy any part of the facilities. Extremists on both sides must have come to understand that it would be suicidal to attack either the dam, which would flood enormous territory on both sides of the border, or the plants, which would cut off everyone's electricity and set both countries back into the nineteenth century. Rational behavior is to be expected when a cease-fire is in place. But even during the war, no irrational steps were taken to damage the facilities.

Even during the periods of heaviest fighting, Abkhaz and Georgian officials and engineers kept electricity generating to the benefit of both sides, without significant interruption. It has been as if the boundary between the dam and the power plants did not exist. At the time of this writing (mid-2000), there is a delicate equilibrium and a shared sense of purpose in joint operation of the plant, as has been the case throughout the entire conflict. Cooperation between the two sides in operating and repairing the complex has been constant, even in the absence of any formal structure of joint management. Clearly there has been a shared and rational notion of why it is important to keep the Inguri complex working, despite the seemingly irrational behavior of both sides that, so far, has blocked a final peace agreement.

We see cooperation taking place among all those with a security interest in this region. Will this cooperation necessarily provide an incentive for peace? Political functionalist thinking suggests that peace can be achieved if adversarial states can collaborate on technical and other nonpolitical matters. The belief is that, over time, this cooperation binds states together and eventually softens and erases political differences. In the process of cooperation, the former adversaries realize that there is more to be gained than lost through peaceful relations. Miriam Lowi (1995, 123) shows the weaknesses in this thinking and maintains that "political conflicts are sometimes so visceral and primordial that they simply cannot be ignored; over the course of their duration, they become an inextricable part of the identities of the parties involved. Under such circumstances, technical collaboration cannot be facilitated; rather it must await political settlement."

A positive example of what Jay Rothman (1991, 107) calls "integrative problem solving and functional cooperation in situations of intense ethnic conflict" is the case of the city of Nicosia, Cyprus, which was divided following war in 1974, leaving a sewage plant on one side of the city and the water on the other. The populations in both halves of the city were left without sewage treatment. Rothman maintains that since neither side had the resources to replace the missing half and were left with no choice but to cooperate, over a number of years the two leaders developed a cooperative plan that was implemented. "This effort," he wrote, "beginning with sewage resources—about as prosaic an effort as one can imagine—has built confidence between some leaders and within their communities that negotiations, at least around issues of urban development, are useful" (Rothman 1991, 107). The joint planning commission that emerged in Cyprus was not a political body and did not address the political future of the city that its members were discussing. Rather, the members agreed to proceed by developing two sets of cooperative plans, depending on whether the eventual political solution would be two separate sovereign states or a binational, bicultural unified island state. As we know, this cooperative effort over sewage treatment has not yet contributed to a long-term resolution of the underlying conflict in Cyprus.

A second explanation for cooperation over the use of the Inguri water resources and the management of the Inguri complex is that international interests, which stand to gain from business as usual, are exerting pressure on both sides. The complex is important not only to the Georgians and Abkhaz, but also to the Russians, who are the mediators at the official negotiations and who command the CIS forces that keep the two conflicting armies separated at the Inguri River. The Russians have an interest in the complex not only because of its military and strategic significance with regard to the peace process, but also because they need the supplemental electricity that the complex provides to their southern territories. Almost every winter, the Russian city of Sochi has needed electricity from the Inguri complex to service its customer demand. It is not surprising, therefore, that the CIS troops guard the complex carefully and attend each bimonthly meeting of the Georgian-Abkhaz managers.

The conflict involves spheres of political and economic influence beyond those of the specific issues for the Georgian and Abkhaz peoples and governments. The region holds strategic significance for both Russia and the West, as it borders Iraq, Iran, and Turkey and is on the cross-roads between the Christian and Islamic worlds. Further, Russia and the West both want access to the world's last major untapped oil reserve. Caspian oil supplies could make Western nations much less reliant on Middle East oil. Extracting and transporting the oil to Western markets could become a four-trillion-dollar industry. One of the preferred routes for an oil pipeline runs from the Caspian Sea region through Georgia to Supsa, a Black Sea port near Abkhazia. Another preferred route passes through Georgia into Turkey.

Other corridors for trade are at stake. Discussions are underway to resume railway and highway transportation through Abkhazia, which will reconnect Russia to important markets in the Southern Caucasus. Preventing further armed conflict and sporadic terrorism, which still continues, is critical to opening up these important transportation routes along the contemporary version of the ancient Silk Road. A settlement is also important to political and economic stabilization and enhancing the prospects for democracy throughout the post-Soviet Caucasus. Georgia is regarded in many Western capitals as having the greatest potential for democracy in the region. For these reasons (and with oil riches always providing an extra incentive), Western nations and several U.S.- and European-based citizen diplomacy efforts, including those of our colleagues at the University of California at Irvine, are playing an increasing role in the Georgian-Abkhaz peace process.

Transnational interests are also displaying an explicit interest in the complex, for example, the European Bank of Reconstruction and Development, which represents the interests of the Western powers currently engaged in the official negotiations. Clearly, the bank and its backers have a tangible stake in the security issues surrounding the complex. To achieve their ends, these institutions are using legal and economic instruments to foster cooperation. In October 1998, the bank resolved to begin financing badly needed reconstruction and repairs of the complex. This decision by the bank has already had an impact on the embattled parties by providing them with incentives to sign agreements ensuring cooperation and security at the site. In the future, the bank may have an even greater, and perhaps decisive, influence on the situation. It will provide a loan of $200 million over a ten-year period to pay for reconstruction work managed by European contractors. For their part, in April 1999, the Abkhaz and Georgian authorities signed an agreement to guarantee the safety of international personnel at the site. The first part of the loan will be for $38 million, to be paid off in three years. Commenting on these developments and the role of outsiders in this effort, an Abkhaz reporter noted that "while the European Community is working on repairs at the Inguri complex, much can change, both in terms of the relationships between the two states, and in terms of the supply of electricity to Abkhazia and Georgia. Most importantly, the complex on the border at all times must be the object of peaceful cooperation" (Tsvizhba 1999, 2).

A third explanation for the cooperation in this complex case study relates to the concept of epistemic communities. The influential epistemic community in this case, comprising Georgian and Abkhaz engineers and managers, is at the same time postmodern (it is a community beyond national divides) and modern (the predominant focal point for this community of engineers is the meaning of water as a modern product and not an ecosystem). Another important element of this epistemic community is that its members share important values that are technocratic and social.

Epistemic communities have been identified elsewhere as important to the creation of cooperative networks (chapters 4, 5, and 9). The Inguri case illustrates that even in the most conflictual circumstances, when national governments dictate the need for cooperation and the top leaders on both sides make the final management decisions, an epistemic community can also play an influential role in decision making. The professional community of hydropower engineers and managers involved in the Inguri complex is a subservient part of the arrangement, but it has a role in creating a favorable work climate for managers and employees at all levels, most of whom have had years of experience working together and have maintained good interpersonal relations. This, in turn, may enhance the efficiency of management and the production of energy.

Abkhaz and Georgian hydraulic scientists and engineers are regarded by their fellow citizens as apolitical; their work was not at issue in the recent armed conflict. The boundaries that separate these epistemic groups from others are professional memberships, not nationalities, and the lines that matter are lines of communication, not sentry stations (chapter 1). One example of this is reflected in a scene we witnessed in Gali. An Abkhaz war veteran was startled to hear a Georgian power engineer express genuine regret that, because of an accident on one of the power lines to the capital of Abkhazia, with its mostly Abkhaz population, the population spent a whole day without electricity. This Abkhaz man was not prepared to encounter the mentality of a professional whose focus was on the provision of electricity and not on the ethnicity of the recipients of his professional service. The engineers themselves express a similar self-image. As one of them said, "This cooperation works because we engineers are peaceable people. If other professionals were like us we could agree on a common space."

As a result, plant personnel believe that they can work gradually, in small ways, toward stabilizing peace in the region. Interviews both with engineers and managers working on the Inguri complex and with politicians reveal their beliefs that this joint management effort is most certainly leading the peace process, not following it. They maintain that the joint management is a model of cooperation, showing that cooperation is possible, and believe that it could also become a model for cooperation in rebuilding railway, communication, and transport ties.

Sociocultural relationships also play an important role in the bonding of this epistemic community. Consultation meetings on the power complex are always accompanied by banquets that, under these circumstances, solidify agreements in a way that is far more binding than a legal contract in U.S. culture. Traditionally, people of the Caucasus do not partake of each other's food if they bear hostility toward one another. So the very act of sitting down to a ceremonial dinner is testimony to a commitment to a relationship. This commitment is reinforced by the etiquette of toast making, one of the main mechanisms of peacemaking in the Caucasus. Throughout a dinner all participants must raise their glasses of wine, vodka, or cognac more than once to articulate a meaningful point, either about the positive qualities of those present, the business of the day, or broader political and cultural issues. An arena for positive discourse is established, since this is a time for all to speak their minds to one another in a positive and encouraging way.

It is also useful to examine factors that are *not* involved in promoting cooperation in water management on the Inguri River. In conditions of war, the centralized authorities on both sides are in complete control, leaving local communities out of formal decision making and financial responsibility for service and security. The broader community in both Georgia and Abkhazia can influence matters only informally. Scientists and academics do not have a voice in governance and decision making, other than in the exploitation of the power plants. Only elite scientists and academics close to government authorities know what is going on in this well-guarded area along the de facto border of Georgia and Abkhazia. In this context, it is unlikely that grassroots opinions will be expressed about environmental or other related conditions that might affect and concern the people on both sides.

The cooperative relationship among Inguri professionals appears to be a unique arrangement in the border region. There are no nongovernmental organizations or religious groups on either side of the river that are building links between the neighboring communities. At present there is more of an informal infrastructure for war, not an infrastructure of cooperation. The concerns of the local population mainly relate to daily physical and economic survival.

As long as no political settlement is implied, the form of Georgian-Abkhaz cooperation represented by the Inguri complex is acceptable both to the Abkhaz and to the Georgians. There is no debate among the public on either side of the conflict about the wisdom of this cooperation, partly because agreements are made quietly, and partly because most people seem to believe that everything is being done for the best. This is in keeping with both the Soviet and the Caucasian hierarchical, patriarchal culture.

The limits of cooperation are narrowly prescribed. Unlike the case of the Yellowstone to Yukon Conservation Initiative presented in chapter 5, environmental issues and ecosystem approaches to problem solving are not a factor guiding cooperation at the Inguri complex. Discourse related to the Inguri River is not framed at all in terms of environmental issues or in relation to concerns about the ecosystem. There are valid ecosystem issues in the Inguri and its watershed, but they are simply not addressed.

Beach erosion has occurred at the mouth of the Inguri River where the water from the canal diversion empties into the Black Sea. Because of dam-related pollution, salmon and sturgeon, which once were abundant in the Inguri River, have all but vanished. Continuing tensions subsequent to the end of the war distract resources and public awareness away from environmental issues, worsening these conditions. Other major dam-related structural problems also have environmental implications. The dam itself has never been filled to capacity because of seepage through cracks and crevices in the rock of the mountain that forms the sides of the reservoir and anchors the dam. One fear is that the fissures may widen within the mountain, or spread to the dam itself, or both. Another concern is the buildup of silt deposits against the dam and in parts of the canal and river structures below the dam. Moreover, there is the potential in the area for earthquakes and volcanic activity. Mt. Elbrus, a mountain located in the geographic region of the dam, is an extinct volcano that some believe has the potential for further eruptions that could damage or destroy the dam and associated hydroelectric facilities.

The engineers who designed the facility and who still participate in its management dispute claims of structural and environmental problems. They maintain that the dam has played a positive role by regulating water flows in the river, preventing frequent flooding and serious downstream damage to crops and homes. They dismiss beach erosion as a continuing concern because of engineering reinforcements along the coastline. Damage to fisheries has been compensated by fish farming on the Kodor River. However, any environmental protection efforts previously undertaken have been discontinued because, at present, there is absolutely no funding for beach erosion control or fish farming.

Environmentalists maintain that it would have been more efficient and sustainable to build smaller hydroelectric plants without dams that would have provided Abkhazia and Georgia with electricity without the negative side-effects. In fact, the plant's output of electricity is considered to be an inferior product for present-day uses because current needs are for base supplies of electricity rather than for the big power surges associated with previous Soviet peaking power demands.

The postmodern ideas that raise concerns about the ecosystem do not appear, however, to trouble the engineers or officials involved in joint management. These professionals take tremendous pride in talking about the dam as being the largest of its kind in the world and assert that the dam has actually improved the environmental conditions in the area. They further maintain that any detrimental effects of the dam and the power plant have been rectified. Only the whispering voices of a few environmentalists contradict the engineers, and they have no impact whatsoever on public opinion or on the formal discourse between Georgian and Abkhaz officials.

Conclusion

In the aftermath of deadly conflict, the joint management of the Inguri River water resources and hydroelectric power complex presents an interesting puzzle. The essentialistic meaning of a security issue does not necessarily ensure active confrontation. This case demonstrates that even when there is a struggle between two states (here, one of them is a de facto state), control of a water resource, critical to state security, does not necessarily result in noncooperative relationships. Cooperation on the Inguri River shows that even under the most problematic conditions, modern and postmodern meanings of water can transcend highly conflictual meanings. This case draws attention to the enduring relevance of sovereignty in dealing with water management, as well as to the importance of geo-spatial configurations, epistemic communities, and cultural habits in contributing to cooperative practices under circumstances not conducive to cooperation.

Major policy decisions about the distribution of electricity from the Inguri plant and about plant operation are made at the highest levels of the governmental structures of Georgia and Abkhazia. In fact, the respective presidents lead the process of decision making on matters regarding the Inguri complex, a circumstance reflecting the significance of the Inguri as an economic and security issue. These are enlightened rational actors, nation-state leaders motivated by the reality of mutual deterrence. In this respect, coerced self-interested cooperation is characteristic of the modern world.

Representing more postmodern elements of the Inguri's operation, the daily technical details are in the hands of an epistemic community of engineers and scientists who operated the complex before the war between the two nation-states broke out. Whatever their private sentiments about the current political situation, they submerge their thinking and loyalties beneath the shared good, coaxing maximum electric power from an underfunded, aging facility with major needs for maintenance.

Other postmodern elements influencing this case are international and transnational forces, such as the European Bank of Reconstruction and Development. These international networks and institutions have a tangible stake in the security issues surrounding the complex and are using legal and economic instruments to foster cooperation and prevent resumption of armed conflict. In this case the most progressive meaning of water (ecosystem management) is not even envisioned.

The lesson from this case study is that, in many parts of the world, modern approaches towards transboundary water management are still useful steps toward progress. Arrangements made between sovereign states are still essential in managing the transboundary Inguri watercourse and its hydroelectric sources. The case demonstrates the fundamental importance of sovereign solutions as bridge mechanisms to local-level sustainable development. Therefore, the modern approach is seen here as only a first, but crucial step in transboundary water management.

References

Carlisle, H. L. 1996. "Hydropolitics in Post-Soviet Central Asia: International Environmental Institutions and Water Resource Control." In *The Political Economy of International Environmental Cooperation.* University of California Institute on Global Conflict and Cooperation. University of California, San Diego.

Golubchikov, Sergei. 1995. "GES 'kavkazskoi' natsional'nosti" [A Hydroelectric Power Station of "Caucasian" Nationality]. *Energiya, Ekonomika, Politika, Ekologia* [Energy, Economics, Politics, Ecology] no. 4: 35–38.

Lederach, John Paul. 1997. *Building Peace: Sustainable Reconciliation in Divided Societies.* Washington, D.C.: United States Institute of Peace.

Lowi, Miriam. 1995. "Rivers of Conflict, Rivers of Peace (Continuity and Transformation: The Modern Middle East)." *Journal of International Affairs* 49, no. 1 (Summer): 123–144.

Paddock, Richard C. 1999. "Freedom Comes at a Steep Price." *Los Angeles Times*, October 1, 1.

Rothman, Jay. 1991. "Conflict Research and Resolution: Cyprus." *Annals of the American Academy of Political and Social Science* 518 (November): 95–108.

Tsvizhba. 1999 "Will Abkhazia Sell Electricity Abroad?" *Respublika Abkhazii* no. 95 (August 19–20): 2.

United Nations Mission on Needs Assessment in Abkhazia, Georgia. 1998. Report by United Nations Development Programme, March. United Nations Development Programme, New York.

Williams, Paul. 1996. "Water Usually Flows Downhill: The Role of Power, Norms, and Domestic Politics in Resolving Transboundary Water-Sharing Conflicts." In *The Political Economy of International Environmental Cooperation*. University of California Institute on Global Conflict and Cooperation. University of California, San Diego.

9

Black Sea Environmental Management: Prospects for New Paradigms in Transitional Contexts

Joseph F. DiMento

Challenges to cooperation in management of transboundary water and related environmental issues are ubiquitous. Barriers to bridge building exist even in the most favorable contexts, as several chapters in this volume illustrate for the North American and western European cases (chapters 5, 6, and 7). However, the opposite is also true. This chapter will argue that even in the most contentious contexts in which forces of nationalism, exclusionary ideologies, and ethnic conflict are separating peoples, cross-boundary networks, discourses on cooperation, institutions that promote that cooperation, and legal regimes that reinforce it are emerging. In fact these forces now exist in the most unlikely situations and are promoting new and additional tools for the analysis and management of water bodies.

This chapter focuses on the Black Sea region. I use "Black Sea region" to denote the six riparian (or littoral) states (and a presently unrecognized former Soviet republic) and the neighboring states that are part of the mammoth watershed of the Black Sea. The riparians are Bulgaria, Georgia (Abkhazia), Romania, the Russian Federation, Ukraine and Turkey. Major rivers that drain into the sea include the Danube, Dneiper, and Don, the second, third, and fourth major European rivers. The only ocean outlet to this gigantic water resource (whose surface area is one fifth the size of its catchment area and which in depth exceeds two kilometers in parts) is the narrow and shallow Bosphorus Channel.

This vast sea presents immense environmental problems, and its environmental management is a formidable task, even in a world of major management challenges. Steps taken thus far include creation of the Black Sea Environmental Programme (BSEP), an example of the international issue-based legal regimes introduced earlier in this volume. In this chapter I describe the evolution of the BSEP, summarize its present characteristics, and discuss them within the context of the evolving new paradigms of water management. Although I focus centrally on the BSEP, my discussion is broader, covering other efforts, less institutionally based, to save the wonders of the Black Sea from further degradation, if not, as some have warned, from ecological death. I conclude that elements of the new orientations to water management have considerable promise in this regional sea context. I argue that these can be integrated or reflected in international law: Formal legal instruments, especially in areas of conflict, are a significant means for orchestrating cooperation on water issues. Law can provide the framework for the contributions of new actors, the structuring of new networks, and the development of shared meanings.

Contrary to some of the perspectives suggested in other chapters in this volume, I contend that international law and its instruments constitute an enterprise that exists alongside and often serves as an enabler of the work of epistemic communities and international nongovernmental groups. I recognize along the way that the law does not uniformly deliver on this promise. I point out in the chapter that within both the existing BSEP and other efforts several obstacles to cooperation exist but so too do conditions favorable to effective environmental management.[1]

To anticipate further, I recommend that to develop cooperative efforts across borders in general, and to address the problems of the Black Sea in particular, careful nurturing of emerging new approaches is needed. These approaches emphasize new actors, new regimes, new methodologies, and participatory values; that is, they further apply the new appreciations of water resource and environmental management.

International Issue-Based Legal Regimes: General

Issue-based legal regimes can be created in several subject areas: trade, the economy, the environment, and combinations of these. By "regime" I mean not only the environmental legal instruments (treaties, protocols, customs, etc.)—of which there were more than 250 at the regional level by the early 1990s—but also the participants in the creation of those laws, the governmental and nongovernmental organizations that attempt to implement them, the rules and norms that they foster, and other groups that act to work within or influence implementation.

In the last three decades the growth of these entities has been remarkably consistent. Now virtually all geographical areas are represented; to name just a few: the Convention on Long-range Transboundary Air Pollution, finalized in Geneva in 1979 (a framework for cooperation among North American and European states to control and reduce pollutants); the Convention for Cooperation in the Protection and Development of the Marine and Coastal Environment of the West and Central African Region, signed at the Abidjan Convention in 1981; the Convention for Protection and Development of the Marine Environment of the Wider Caribbean Region, enacted in 1983; the Convention on the Protection and Use of Transboundary Watercourses and International Lakes, formalized in Helsinki in 1992 (a framework treaty aimed at eliminating contradictions with evolving principles of international environmental law such as the precautionary principle, the polluter-pays principle, and the principle of consideration for future generations); and the North American Agreement on Environmental Cooperation signed by Mexico, Canada, and the United States in 1994 (chapter 7).

By some criteria, the most advanced regional efforts are those of the European Union (in general see chapter 4). The Single European Act of 1987 built environmental policy into the Treaty of Rome. The Treaty of the European Union of 1992 enhanced the union's authorities and allowed for the use of majority voting on environmental legislation. It also introduced the concept of sustainable growth respectful of the environment. The Maastricht Treaty, adopted in 1994, describes a comprehensive agenda for sustainable, noninflationary, environmentally sensitive growth.

Several characteristics are common to international issue-based legal regimes. Decisions are made at a supranational level that have a binding effect on the national members. Some also allow member states, or even citizens of member states, to bring actions in a judicial or other dispute resolution forum to address the alleged noncomplying action of another member state. Sanctions or some other form of redress may be available if a member state is found in violation of, or not in compliance with, the instrument's requirements. Some regimes foster the participation of previously nontraditional actors in international relations: nongovernmental organizations (NGOs), multinational corporations and industry associations, and multilateral development banks. And some require application of new methodologies in management of environmentally relevant projects, such as environmental impact assessment (EIA).

International law, however, has only recently developed mechanisms for protecting and managing large marine ecosystems (Belsky 1985). In 1973 the United Nations Environment Programme (UNEP) Governing Council declared the regional seas to be an area of special priority. Later it sponsored twenty-three treaties through its Regional Seas Program. The Barcelona Convention on the Protection of the Mediterranean Sea against Pollution of 1976 was the first such treaty, and it set a pattern for these conventions. A framework instrument is entered into that allows member states to adopt jointly or severally appropriate measures to prevent, reduce, and control pollution in general and from various specified sources. Members also agree to cooperate in monitoring and in addressing critical problems. The Mediterranean program progressed with more specific protocols: on the protection of the sea against pollution from land-based sources, signed in Athens in 1980, and on specially protected areas, finalized in Geneva in 1982. In all there are now more than forty regional marine environmental treaties and protocols (Biermann 1998).

The actual positive environmental results of the regional seas efforts have been mixed and rather modest, in part because of their failure to reflect more fully modern orientations to water management (*Declaration* 1992). For example, the objectives of the Mediterranean Action Plan have been reconfirmed but limitations on its effects have been recognized (*Barcelona Resolution* 1995). Thus law remains a strong promise for many regions, one that can be more fully realized if elements of the new paradigms are consistently integrated. Furthermore, in numerous other regions, a shift to these analysis and management methods is essential to prevent further environmental devastation. Many of these problem areas fall within or near the former Soviet Union. Among these are the Kura, Sraks, and Saur rivers in the Caucasus; the Caspian Sea, involving Azerbaijan, Russia, Kazakhstan, Turkmenistan, and Iran;

the Aral Sea (Kirghizstan, Tadjikistan, Afghanistan, Turkmenistan, and Uzbekistan) (Vinogradov nd); and the present case, the Black Sea. But the challenges are distributed across all regions (Milich and Varady 1998).

Background and Need for New Black Sea Environmental Management

Although scientists debate and analyze exactly how serious the situation is, pollution and ecological degradation of the Black Sea is on nearly everyone's list of major environmental problems in the world. Under the Soviet system, a large number of specialists in all areas of relevance to water body management studied and worked on Black Sea environmental problems. However, the links between their work and official decision making on environmental matters were not strong. As a Georgian retrospective summarized: "National environmental legislation was often based upon objectives and standards which were too strict to be enforced or were not linked to effective economic instruments such as fines or permit charges. As a result of years of isolation, many institutions lacked the modern equipment and know-how necessary to face the challenge of providing reliable information on the state of the environment itself" ("State of the Environment Georgia" 1996). Problems were even greater than this summary suggests and involved lack of coordination among the Soviet states and their neighbors, including Turkey. Lack of public participation, absence of transparency of decision making, and absence of other factors that promote implementation, such as a modern regulatory approach, technical assistance, and adequate funding, made the challenge even more difficult to meet.

The environmental problem in the Black Sea is multifaceted, ranging from loss of landscape to extinction of species. Scientists have used various terms for the ecological health of the Black Sea, whose ecosystem "has changed irreversibly" (Global Environment Facility 1997, 139). By the early 1990s, terms like "dead," "close to collapse," "unholy mess," and "ecological disaster" were common descriptors of the status of this giant, beautiful natural resource, and the preceding three decades were generally acknowledged to be a period of persistent environmental degradation.

Pollution's effects on the Black Sea are widespread, negatively affecting recreation, tourism, biodiversity, fishing, and water quality. The destruction of the fish species in the sea is generally considered one of the greatest ecological catastrophes of the modern era.

The Black Sea's directly bordering, littoral countries include Turkey and republics formerly under the control of the Soviet Union whose cleanup technologies, monitoring stations, and environmental laboratories are in great disrepair. As the watershed area (the collection basin) for more than thirty rivers, the sea receives the effluents of 160 million people from seventeen nations, one third of Europe. It is also polluted by oil and the radiation fallout from the accident at Chernobyl and, by some accounts, from heavy metals, including chrome, copper, mercury, lead, and zinc (Sampson 1995).[2]

The quantity of organic matter from the rivers that feed the Black Sea is so great that the bacteria in the sea cannot decompose it completely. In the Bosporus Strait alone the untreated sewage of ten million people is dumped regularly; this represents only about 6% of the total pollutants received into the Black Sea (Sampson 1996). Nor can dissolved oxygen complete the process. Organic material strips oxygen from sulfate ions and creates hydrogen sulfide, a toxic gas. The Black Sea "is the single largest reservoir of hydrogen sulfide and the biggest natural anoxic basin in the world. To a depth of 150–200 meters, the sea is teeming with life, but below that level, the water is 'anoxic' or 'dead.' There is no oxygen below this level, so no fish, shellfish or bacteria live there" (GlobaLearn 1996).

Biodiversity loss was and is a major problem resulting from eutrophication (overfertilization of a water body with nitrogen and phosphorous compounds, in the Black Sea case from fertilizers and urban and industrial sewage), "clearly the main ecological concern in the Black Sea," in the words of the *Black Sea Transboundary Diagnostic Analysis* (Global Environmental Facility 1997, 70). The resulting overproduction of phytoplankton and reduced sea grass and algae result in concomitant loss of crustaceans, fish, and mollusks.

Another cause of biodiversity loss in the Black Sea is the introduction of exotic species. For example, the *Mnemiopsis leidyi* was introduced into the region by accident from the eastern seaboard of America in the ballast water of a ship. This jellyfish-like species, which consumes fish larvae and tiny animals that small fish feed on, reached a mass of 900 million tons in the 1980s, which is ten times the annual fish harvest worldwide, pushing many fish species to extinction. The fish catch in the sea degenerated to a quarter million tons in 1991 from a total of 850,000 tons less than a decade earlier.

Tanker and operational accidents in the sea have been sources of oil pollution, as has the direct dumping of solid waste into the sea or onto wetlands. The polluting impacts of rapid oil industry development (1,500 tankers pass through the Bosporus Strait in each direction annually), sedimentation impacts, beach erosion, and the overall absence of coastal zone conservation are also strongly felt. Sturgeon and shad cannot run upstream to breed because of damming of the big rivers that drain into the sea.

Reports differ on the overall total number of species affected by the various sources of pollution. One estimate is that two wildlife species have disappeared and habitat has been degraded for more than 44,000 species of marine animals. Other reports conclude that the number of fish species in the sea dropped from around twenty-five to between three and five in the ten-year period from 1986 (when the sea had five times the fish production of the Mediterranean) to 1996. (Conclusions regarding the status of species differ, and the 1997 BSEP *Annual Report* (UNDP et al. 1997) states that thirty-three species exist in the Black Sea, with four species providing 80.4% of the total catch.) The *Black Sea Transboundary Diagnostic Analysis* (Global Environmental Facility 1997) describes the problem of the loss or imminent loss of endangered species in the sea and its wetlands as affecting bottom-plant communities, mollusks, bottom crustaceans, dolphins (5–10% of the reference level remain) and monk seal (few specimens remain), among others. Giant sturgeon are endangered, other sturgeon species are depleted, turbot are seriously depleted, blue mussels are in serious decline, native gray/golden mullets are depleted, and rays are likely depleted, as are red mullet.

International Issue-Based Legal Regimes: The Black Sea Case

The BSEP, developed under the auspices of the UNEP and the Global Environmental Facility, is one response to the sea's degradation. The Programme was established in the early 1990s and modeled on the

Barcelona Convention. Bulgaria, Georgia, Romania, the Russian Federation, Ukraine, and Turkey signed the Convention for the Protection of the Black Sea against Pollution (the "Bucharest Convention") in April 1992 in Bucharest, and it was rapidly ratified by each country by spring 1994. A Ministerial Declaration on the Protection of the Black Sea followed; it was signed in April 1993 in Odessa, Ukraine. Modeled on the Agenda for the Twenty-First Century adopted at the Rio Summit of World Heads of State in 1992, it declared among other goals "protection, preservation and, where necessary, rehabilitation of the marine environment and the sustainable management of the Black Sea" and that countries should elaborate and implement national integrated management policies, including legislative measures and economic instruments, to ensure sustainable development. It encourages public participation (including that of NGOs), the precautionary principle, use of economic incentives, environmental impact assessment, environmental accounting, and coordination of regional activities.

The Bucharest Convention entered into force on January 15, 1994. Other affiliated international legal instruments that make up the BSEP regime include the Protocol on Protection of the Black Sea Marine Environment against Pollution from Land-Based Sources (April 21, 1992), the Protocol on Cooperation in Combating Pollution of the Black Sea Marine Environment by Oil and Other Harmful Substances in Emergency Situations (April 21, 1992), and the Protocol on the Protection of the Black Sea Marine Environment against Pollution by Dumping (not yet in force, as of May 2000).

Initially, funding for the BSEP was provided by the Global Environmental Facility, the European Union, and Austria, Canada, Japan, the Netherlands, Norway, and Switzerland. Funding presently comes from the UNEP and is to come from the member countries (personal communication, BSEP Programme Coordination staff member, August 28, 1998; see also UNDP et al. 1997.) The UNEP's role in BSEP history is consistent with this international governmental organization's (IGO's) mission, which is to "provide leadership and encourage partnerships in caring for the environment by inspiring, informing and enabling nations and people to improve their quality of life without compromising that of future generations" (http://www.unep.org/Documents/Default.asp).

The Programme Coordination Unit (PCU) of the BSEP was placed in Istanbul. It was replaced by a Project Implementation Unit comanaged by the United Nations Development Programme (UNDP) in spring 1998 in the hope that this implementation unit would be a precursor to a secretariat to be financed by the member countries.

The BSEP held its first meeting in Bulgaria in 1993. Three objectives were highlighted: improving Black Sea country capacity to assess and manage the environment, supporting the development and implementation of new environmental policies and law, and promoting sound environmental investments. Activity centers to be hosted by the individual Black Sea countries were created: the Center for Biodiversity and the Black Sea Ecology and Fishery Institute in Georgia; the Center for Environmental and Safety Aspects of Shipping in Bulgaria; the Center for Integrated Coastal Zone Management in Russia; the Center for Fisheries and Other Marine Living Resources in Romania; the Center for Pollution from Land Based Sources in Turkey; and the Center for Monitoring and Assessment in the Ukraine.

Lawrence Mee, an academic who wished to bring to the program new water management perspectives, became the program's first coordinator in 1993. He focused on the need to incorporate public participation in BSEP activities and was unusually candid in his assessment of the strengths and weaknesses of the contributions made to the Black Sea improvement by NGOs, nation-states, and others. Extremely knowledgeable about the region and technically capable, he guided the program through its early years. In 1997, after five years with the BSEP, he decided to leave the organization to return to his role as a professor.[3]

The Eighteenth Session of the Intergovernmental Oceanic Commission (IOC) United Nations Educational, Scientific, and Cultural Organization (UNESCO) Assembly in 1995 reported some progress in Black Sea management: an agreement to approve a Black Sea Regional Programme in Marine Sciences and Services. Before this agreement was reached, an intergovernmental meeting in which all Black Sea coastal states participated was held in Paris in June 1995. The United Nations, through its Office for Project Services (UNOPS), which assists developing nations and countries in transition, has provided important services to the BSEP since the early 1990s through management of scientific institutions that address emergency response, pollution monitoring, biodiversity protection, and other foci. It also managed the BSEP's PCU in Istanbul. UN efforts also addressed the creation of interagency agreements for most UN agencies that work in the arena of the marine environment (Food and Agriculture Organization, International Maritime Organization, World Health Organization, UNESCO, and others) and the upgrading of local laboratories (opSEA 1998).

Global IGOs and the European Union continue to make contributions to the Black Sea efforts. The union has provided, through its multinational program for the development of commerce (PHARE) and Technical Assistance for the Commonwealth of Independent States (TACIS) programs, millions of ECUs and consulting assistance from Britain, Denmark, Holland, and Spain, as well as demonstration projects in aquaculture, support activities for creating a Black Sea Environmental Fund, introduction of integrated licensing procedures for regional inspectorates, development of a regional monitoring network, coastal zone management, rehabilitation of the dilapidating dolphinarium in Batumi, Georgia, and public awareness and environmental education activities.

In October 1996, the Black Sea border countries signed a Strategic Action Plan. The plan's preamble reaffirms the commitment of the members states to the rehabilitation and protection of the Black Sea and the sustainable development of its resources. One element of the short plan, which the BSEP describes as a flexible document responsive to contingencies, sets out principles seen as the basis of international cooperation. Among the fifteen background principles are several that are also elements of new perspectives on water management: introduction of compulsory EIA, environmental audits, and strategic EIAs with harmonized criteria; increased attention to public participation based on a comprehensive package including local authorities, NGOs, the private sector, regional environmental centers, and schools; and enhanced transparency through rights of access to information and improved public awareness.[4]

The Programme has had some serious problems with implementation, including very slow realization by member states of the commitment to modest funding. Upon his departure, Mee gave the Programme an extraordinarily candid evaluation:[5]

The truth ... is that very little has been done to fulfil the initial commitment made to the people of the Black Sea countries when their six legislative assemblies ratified the convention in 1993.... [D]ecisions taken through democratic processes have been disregarded and political momentum has been lost.... This scenario is a depressing one. (UNDP et al. 1997)

Assessment: Physical Parameters

As of 1997, a BSEP report could provide a somewhat more encouraging perspective of the physical status of the Sea. According to the *Strategic Action Plan of 1996* (http://www.blackseaweb.net/action/content/html), the *Black Sea Transboundary Diagnostic Analysis* "clearly demonstrates that the Black Sea environment can still be restored and protected." The plan concluded that "environmental monitoring conducted over the past 4–5 years... reflects perceptible and continued improvements in the state of some localised components of the Black Sea ecosystem." Furthermore there are reports that, although still a plague, the *Mnemiopsis* is in decline.

Assessment: New Regimes in Transitional Contexts

In this section I discuss factors that have been linked to innovative approaches to transboundary cooperation for the Black Sea. Some are clearly present in the Black Sea context, not only in relationship to the specific institutional entity (the BSEP) that is the center of this analysis, but also in the larger context of institutional initiatives. The analysis of other factors suggests, at least for now, comparatively modest movement toward international water cooperation for the sea. But unpredictable historical events can dramatically modify modest expectations (as the fall of the Berlin Wall and the collapse of the Soviet Union remind us).

Major Barriers to Cooperation

Exclusionary Ideologies   Perhaps the greatest barrier to cooperation in the Black Sea region is the emergence of two types of inward looking movements in the region, nationalism and religious fundamentalism. These movements reflect values that are inconsistent with the new paradigms of interaction. They emphasize ideologies that associate cooperation with the diminution or destruction of important national values or strict exclusionary fundamentalist understandings of religious beliefs. At the beginning of the twenty-first century, both of these movements may again be in the ascendancy in both Turkey (Sampson 1995) and the former Soviet states. Even small percentages of actors with ties to such movements can pose grave obstacles for communitarian and inclusive approaches to any type of international policy, water management being only one case.

Nationalist movements may perceive environmental degradation resulting from the activities of neighbors as not being improved by their own efforts. Fundamentalist movements may be little concerned with the environment but rather emphasize an overarching concern with limiting interactions, including cooperation with societies of different beliefs, which engender exposure to unwanted values and lifestyles.

Access to Place-Based Perspectives and Information   The BSEP, organizationally and at least theoretically, has access to a variety of perspectives on the water resource that it seeks to manage. Of future import will be the actual access to perspectives from "on the ground" users of the sea who have valuable knowledge about ecosystem perturbances in and around the Black Sea region. Communication of their knowledge and experience to international organizations is, in many cases, limited or nonexistent. Also problematic has been the absence of access to information, data, and perspectives from Abkhazia, an entity not recognized by the international community. Its hostilities with Georgia, detailed in chapter 8, virtually eliminated its communication with international organizations, except in isolated instances. The conflict between the two countries has thus created gaps in information available to the BSEP. Data about conditions in the area of Abkhazia and information from its scientists and officials are lacking. Scientists cannot get in or out of Abkhazia; information flows have ceased. However, BSEP personnel have expressed a willingness to seek means to make use of data from the area. Officially governments may not be receptive to recognition of Abkhazia, but on some matters of concern to Black Sea management, data from Abkhazia are sorely needed. These data may be accepted if channeled through one set of new actors, third-party neutrals, such as university-sponsored peace groups and the UN volunteers until regular relations are restored (DiMento 1998). These new actors are most effective in channeling information when viewed as valuing international cooperation but open to interpretations by the parties themselves as to the nature of the interactions, including the timing of data sharing and the type and quantity of data. Furthermore, these new actors are most useful if committed to transboundary facilitation in the mid- to long run and if conversant with a wide range of conflict resolution techniques.

Promoters of Cooperation

These and other obstacles to regional cooperation, such as the limited infrastructure for communicating across national boundaries (Sampson 1995) even when the intention is established, should not be underestimated. Yet several other countervailing forces in the region make prospects for the refinement and implementation of new regimes more promising. This section inventories those factors while recognizing several constraints on their full positive influence.

Scientific Findings on the Nature and Scope of the Environmental Challenge   There is some scientific debate about several parameters of the Black Sea environmental status, including the extent of human contribution to the hydrogen sulfide cycles and amenability to mid-scale interventions. And there are several areas of scientific uncertainty around the assessments made of the sea, including, for example, those of the discharge of chemical and microbiological contamination in coastal and marine areas. Furthermore, only in recent years has there been movement toward the standardization of the formerly differing scientific protocols and methodologies for scientific investigation, even within the participating nations (Sampson 1995). However, most actual and potential participants in the analytical processes recognize the search for better data, more precise models, better equipment to test models, and basic science to underpin the models as an opportunity for cooperation. Although present economic conditions hinder the realization of the full potentials of the scientific communities in the former Soviet states, it is commonly recognized that the region has a rich resource of scientific expertise. What's more, the international community of environmentally progressive nation-states and UN organizations has targeted the Black Sea as an area deserving of major contributions of technical expertise and funding.

Shared Perspectives   The different levels of development and the number of cultures, religions, languages and views of history represented in the peoples of the Black Sea basin, from one perspective, make the sharing of outlooks on cooperative activities to improve and maintain the environmental quality of the Sea appear daunting. Yet precedents exist for establishing cooperation in similar circumstances (Kliot and Shmueli 1997; Milich and Varady 1998). Furthermore the Black Sea has had immense historical importance to each of the littoral members. To witness degradation, if not collapse, in a matter of decades of a world resource that has provided transportation for emperors, food for nations, recreation for all, and spiritual meaning to many suggests that common understandings on this environmental challenge may be more readily achieved than on other matters of international policy. In the Black Sea context, returning to these historical meanings may be the goal for promoting cooperation (Ascherson 1995). For others, however, a broader meaning of water needs to be learned, whether through environmental education, environmental mediation or other methodologies: water as an aesthetic treasure, water as the source of sustainable life, water as a channel to link people rather than divide them, water as a focal point for building regional communities in the post-Soviet era, water as source, not as sewer. Again Mee's assessment is, however, balancing if not sobering: "I am always amazed at the amount of trash on Black Sea beaches, in national parks and by the roadside. Sometimes, I find protected species in the fish market. Beautiful chestnut trees are burned for firewood.... In the Black Sea region, an entirely new approach is needed if real changes are to occur in environmental awareness" (UNDP et al. 1997, iii).

In addition, there are an increasing number of common interests, shared by each of the riparians, in economic development. The relative success of the Black Sea Economic Program, a parallel regional effort, is a vivid reminder that economic interest may be a most effective vehicle to promote cooperation. This conclusion is not incompatible with environmental protection, either in general, or in this case. (Although recent improvements in the ecological health of the Black Sea have been correlated with reduced economic activity in the area, economic recovery can also be managed with a focus on environmental protection. Economic gain can coexist with environmentally oriented management of the Sea. Only in the narrowest, most short-term perspective are economic benefit and environmental enhancement incompatible.)[6] In the Black Sea case, failure to provide assurance of a relatively clean and safe environment can further damage the already threatened tourist economy. Absence of sound regional environmental policies may also influence IGO and private international funding of major economic development projects, such as oil pipelines. Furthermore, a sustainable fisheries industry, including aquaculture, is a direct source of economic gain for the area. The *Black Sea Transboundary Diagnostic Analysis* reported that, in direct relation to overcapitalization of fisheries and overexploitation, annual values for the fishery declined by tens, perhaps hundreds, of millions of dollars between the 1980s and 1990s (Global Environmental Facility 1997).[7]

Shared values, however, do not always convert to shared understandings of what is to be done. Some observers of the Sea, locals who reside and live by its bounty, consider human input into the problem to be trivial. They feel the Sea will survive without the man-made efforts of international movements and bodies:

Everyone is quick to explain that humanity is not to blame for the anoxia. These dead depths have existed for thousands of years because of the fertile rivers flowing into the closed sea. This disaster, if one calls it that, was made by nature itself. A shipbuilder (an informant to one study on the sea) says, "Of course I know about the gas. It was here long before humans were." Dinko [another informant] explains that the anoxia is not because of pollution. "The anoxic layer keeps rising. That's the problem. What we have must be kept clean." The shipbuilder says the levels are not rising, as some scientists claim ... the levels ... fluctuate up and down. There is a concern that the anoxic levels at the bottom of the sea exist in a delicate balance of pressures and densities. A change in these pressures could cause a "turnover" where these layers come bursting and kill life in the upper layers of the sea. The shipbuilder claims that he has read in newspapers of incidents when the gases bubbled to the surface and no "incident" occurred. "We pollute the sea. it doesn't pollute us," he concludes. (GlobaLearn 1996)

Shared Perspectives and Law   Numerous legal commentators have voiced disappointment with the principles of water management that treaties have adopted or leading international law organizations have recommended. Of the more than 2,000 treaties that have addressed transboundary waters, many rely on the principle of equitable utilization (apportionment of benefits and uses of water in an equitable way) (Fuentes 1997). Stated theoretically, the principle is compatible with new paradigms; the discrepancy has arisen in interpretations and, as with much international law, implementation. But the absence of new paradigmatic law in other international instruments does not preclude any single region from adopting new perspectives. The UNEP and the European Union are both striving to add new meaningful dimensions to the principles of water law.

International law, when evaluated as a tool for promoting effective management strategies for international waters, is an almost empty suit that can be filled in various ways. It can codify the new shared perspectives on high-quality water: as habitat for multiple species; as a transportation corridor for well-managed natural resources, including oil; as the source of recreational and esthetic benefits to multiple peoples; and as a spiritual good. Even in situations in which it sets rather vague standards, the law can increase the probability of cooperation by requiring member states to negotiate rather than to use more adversarial dispute resolution processes and by encouraging access to perspectives from NGOs and other private sources (Benvenisti 1996). New orientations can achieve great significance if they are realized through international legal means and maintained through international legal mechanisms. It may be overstatement to conclude, as Dellapenna (1989) does (quoted in Nollkaemper 1996, 43), that "there can be no answer without international law." Indeed, as with straddling stocks (Davies and Redgwell 1997) and some customary and proposed treaty international law on riparian use generally (Nollkaemper 1996), legal experts can actually impede the movement toward new management solutions.

Nonetheless, international law can be instrumental in resolving conflicts of uses of water sources. History has shown that without a sufficiently detailed legal structure accepted by riparian states, the resolution of conflicts is all too often prone to failure. Accepted legal entitlements grant legitimacy to some claims and deny it to others. They help to shape a common understanding of riparian states' entitlements that can be the basis of agreed solutions to conflicts of uses. International law in itself, however, will not provide solutions to such conflicts (Nollkaemper 1996, 43).

The Black Sea regime recognizes new principles of international environmental law. For example, in the *Strategic Action Plan*, Section II, "The Basis for Cooperative Action," states that the concept of sustainable development is to be applied and that mechanisms for establishing decision making processes as open to interested parties are to be fostered (*Strategic Action Plan* 1996).

New Actors   With significant caveats about nationalist and fundamentalist movements, it seems clear that the Black Sea region is rapidly being enriched by new actors trying to claim ground in both domestic and international policy and working to construct bridges where walls recently existed. Numerous new NGOs have emerged in the BSEP nations. For example, Georgia alone, with a population of only around 5.5 million, has some 200 registered NGOs, of which a large percentage focus on the environment. Black Sea–region NGOs provide perspectives and information to the environmental decision makers in the region, as they do worldwide in the case study presented in chapter 5. They also act, thus far in a very circumscribed manner, as conduits for information that could not be made available through official channels; this includes providing environmental impact assessments that might otherwise be deemphasized because of the national economic growth objectives of the parties (oil exploitation and transport, for example) and, in the present case more specifically, providing knowledge from Abkhazia.

Another caveat beyond those concerning nationalist and fundamentalist movements exists about these new actors. Sometimes NGO presence does not signal the addition of fresh and necessary perspectives. Rather, some NGOs exist for exclusionary or nationally competitive purposes. Others, especially in regions with immense competition for limited external resources, work to promote mainly their own continuation. Personnel within these institutions seek to substitute through the NGOs for their former academic or other salaries, with the goal in the NGO of new analyses either displaced or lacking from the outset. In other words, their commitment to other issues, including those related to the environment, is secondary to their self-advancement goals. As former BSEP head Mee stated within the context of international organizations:

Where are the Black Sea NGOs in all of this? Sadly, their role is often as weak as the government agencies. In many cases they are disconnected from the "grass roots" of society and have become special interest "clubs" of individuals who huddle together shielding themselves from the outside world.... It sometimes surprises me ... that so much energy is put into meetings rather than "hands on" activities." (UNDP et al. 1997, iii)

Also, as to other new actors, "[o]rdinary people are now ready to fight for the Black Sea" (Ascherson nd). To be sure, they face a perception that "their governments, struggling with the colossal costs of the transformation to capitalism, cannot spare cash for the environment" (Ascherson 1996). This is a refrain that describes the disconnect in many areas between what surveys show to be the priority put on local natural environment and global resources and the actions taken by governments, even those democratically elected. The notion that the environment must wait may be one held more by elites than people close to the destruction of the environmental resource: The latter recognize that the loss of their livelihood and environmental destruction are simultaneous. To some extent, removing obstacles to participation by new actors may be more effective for environmental goals in transboundary settings than creating official new government structures (UNDP et al. 1997). Put positively, removing obstacles means allowing citizen input on important decisions that affect the environment through public participation, comment on environmental impact, and access to forums in which decisions are adjudicated. Most formally, under evolving national and transboundary legal systems, this may mean granting legal standing to parties, individuals, and NGOs, which are not formally recognized in the decision-making structures of some of the parties.[8]

These new actors are being assisted throughout the region by established institutions, which have resources that can make the efforts of new players more effective. These include (as already noted) the important, if spotted, contributions of major IGOs such as UNEP and the Global Environmental Facility (GEF) (Caldwell 1990). But the nongovernmental sector also offers major resources. The American Bar Association has created a Central and East European Law Initiative that has provided a number of Environmental Advocacy Centers; these assist NGOs and citizens in enforcing rights and in furthering the goals of democracies. The initiative has also worked with other groups in the area, including the Soros Foundation's Institute for Constitutional and Legislative Policy, to train the public to participate more effectively in decisions that have environmental impacts. Numerous foundations have made the support of embryonic democracy and public participation advocates, environmental NGOs, and what have come to be known as civil society movements a funding priority.

New Actors/Shared Perspectives/"Ecoboat": Spiritual Values and the Sea It is not unrealistic to conclude that new and innovative coalitions whose objectives are transboundary water cooperation will continue to form. For example, symposiums have been held under the auspices of the Ecumenical Patriarch of Constantinople and the European Commission to address the ways religion and science can cooperate for environmental protection. One of the symposia took place on a ship on the Black Sea. It involved 350 participants, including scientists, major leaders of the orthodox churches of the region (the first time they had met since Soviet rule ended), other religious leaders, politicians, and philosophers. The ecumenical initiative supports "the establishment of an international institute of environmental ethics which would bring together representatives of religion, science and the humanities charged with the responsibility of advising international and national leaders on questions of environmental ethics, which would eventually be invested with the power of national and international law" (Pergamon 1997: html:// www.patriarchate.org/patriarchate/visit/html/enviro_index.html).

As the symposium participants concluded,... environmental problems such as pollution are reflections of many general global issues affecting humanity as a whole. Revolutionary changes in world affairs, spurred on by the globalization of markets, demonstrate that ecological, development and societal problems cannot be solved by fragmented and sectoral initiatives alone. Their solution requires ... multidisciplinary interlinked and comprehensive approaches ... vision and long-term commitments. Peace, economy, environment, social justice and democracy are integral parts of the whole. (Pergamon 1997: http:// www.patriarchate.org/patriarchate/visit/html/enviro_index.html)

Before the cruise that focused on the Black Sea, an earlier voyage-symposium addressed the destruction of the environment in general.

Epistemic Communities ... of Scientists and Others   Work in environmental policy and in practical peacemaking has identified small groups of epistemic communities that can complement the recognized groups that have traditionally worked across areas of science policy in the former Soviet Union. Epistemic communities are communities without borders—of scientists, lawyers, engineers, or other specialists. They share core beliefs and understandings and have strong alignments with a particular objective that transcends their concern with affiliation with a political jurisdiction or position. To be sure, some recent discourse is less sanguine about the objectivity and progressiveness of these communities, but generally epistemic groups have been described as neither ethnically nor politically bounded. Their work goes on despite political hostilities among their home countries and provides participants with opportunities to interact with professional colleagues who do not bear the mantle of "enemy." Epistemic communities have been valuable, for example, in the progress—measured, if modest—in the development of the Mediterranean Action Plan, a multinational effort to provide research results and cooperative problem solving to address problems of pollution in that vast sea (Haas 1990).

Other Factors
Several other critical factors will influence movement toward a new Black Sea regime.

Dispute Resolution Mechanisms   Closely linked to the involvement of new actors is the choice of dispute resolution methods to which those actors will be invited to contribute. The BSEP has not developed such mechanisms, but other opportunities for their development and use exist. In addition to those provided by third-party neutrals to parties that seek resolution of conflict, access to European forums may be available. In this regard, a potential legal action by scientists from the Black Sea nations against Austria and Germany is illuminating. The action would challenge nitrogen discharges by the two countries into the Danube—more than 200 tons a year, which is 35% of the total receipt of nitrogen into the Black Sea. The discharges may violate the European Union's directives on wastewater and nitrogen, and the scientists' highlighting of them would bring great embarrassment to nations that take pride in pursuing strong environmental protection policies within their own borders and in other international contexts. This technique—using embarrassment or the threat of embarrassment to alter behavior—has been used extensively in the case of the Yellowstone to Yukon Conservation Iniative, as described in chapter 5.

The scientists reportedly made the decision to pursue this legal action against Austria and Germany on the ecovoyage described above. There are conflicting views of what actually was proposed in terms of legal action (UNDP et al. 1997), and actions, if pursued precipitously, can counter embryonic cooperative activities. Mee wrote that the actual loads of nutrients disgorged by the coastal countries suggest that "while not material for a law suit, the responsibility is demonstrated to be a shared one" (UNDP et al. 1997). Nonetheless, in some settings, resorting to more formal dispute resolution mechanisms may be a means of calling the attention of elites to the voices of new actors.

The availability of forums that allow NGOs and private citizens to directly confront environmentally insensitive actions taken by governments is an element of the new paradigm that has had limited application in international contexts. Brunnee and Toope (1997, 47) conclude that "despite the numerous dispute settlement provisions included in international environmental treaties, these mechanisms are not widely employed. Dispute avoidance schemes linked to river commissions, such as consultation mechanisms and prior notification rules, have proven useful, but most third-party dispute settlement processes remain unused." However, if properly designed, forums that bring together in relatively nonadversarial ways those with direct interests in the outcomes of Black Sea environmental management may be the most direct means to reach commonly held transboundary environmental goals.

Organizational Characteristics of the Embryonic BSEP Itself   Parts of a regime—organizations such as secretariats and, in the Black Sea case, the Istanbul PCU—ultimately need to convert understandings of water and its management into concrete actions. As for the BSEP itself, several characteristics will be important. Among those generally relevant to international environmental organizations are flexibility in responding to new understandings of the nature of the Sea's problems; perceptions of the BSEP as legitimate among those whose behaviors must be influenced for proper management (oil companies, farmers along the feeding rivers, ecotourism and fishing enterprises, municipalities); openness to public input and transparency of decision making; and operational capability of implementing ideas for funding management programs, such as the proposed fines on polluting oil tankers. Each of these is a characteristic of rather sophisticated organizational development; combined, they pose a considerable challenge to the embryonic BSEP.

Choice of funding mechanisms is critical to the promotion of international cooperation. In this stage of BSEP development, imposition of fines on violating oil tankers, as the BSEP is considering doing, might hinder some elements of cooperation (DiMento and Doughman 1998). Ecosystem regime development, as Brunnee and Toope have argued (1997, 32), may best be fostered by focusing on implementation of principles rather than enforcement, "which tends to view normativity in more static terms as the direct application of fixed rules through force or the threat of force." Nonetheless, some "polluter-pays" element appears essential in a region where financial resources are limited. Furthermore, such fines may not interfere with cooperation if observers view fines as equitably imposed on entities that are benefiting greatly from access to the natural resources of the region.[9]

Political Stock/Leader Support for Cooperation   The fragility or strength of the BSEP also depends in significant part on the commitment of the leaders of its riparian members. These leaders are involved in the kind of two-level game described in chapter 1: One level is international, the other, domestic, with its constraints on ability to fully cooperate across national boundaries. Support at the international level is measured partly in financial contributions (which have been slow in coming, even at the very modest requested levels), but not only in that manner.

Public recognition of an international organization's successes, active participation at its meetings, coordination of its activities with individual nation-state management efforts, and attention to staffing decisions all demonstrate how political leaders value an international entity: Is it something merely to be tolerated or an innovation to be nourished and assisted actively? The economic and political challenges in the Black Sea region, with problems of currency devaluation, ethnic conflicts, and social program priority setting, serve as significant but not insurmountable obstacles to leaders' attention to water issues. So too does the extreme weakness of the environmental sector in each of the Black Sea governments (UNDP et al. 1997, ii). Significantly, some MARPOL [The 1973 International Convention for the Prevention of Pollution from Ships and its 1978 Protocol are known as MARPOL (short for *Marine Pollution*)] provisions and those of other agreements related to oil pollution management have not been implemented in the last several years for numerous reasons ranging from equipment failures to difficulties in monitoring and communication on the high seas. Politically, in theory, active support of environmental protection is attractive at this time in the region, both to please emerging green domestic constituencies and for additional, regional motives such as to gain admission to the European Union and access to the Global Environmental Facility and other international environmental funds.

Recognizing Existing Linkages   Study of existing cooperative mechanisms provides lessons for new attempts and suggests that if the conditions are adequate, cooperative efforts can be successful. One remarkable example is the cooperation between Georgians and Abkhazians in joint management of a hydroelectric power plant on the Inguri River (chapter 8). Certainly other, less dramatic examples are available as models for the Black Sea participants: the North American Free Trade Agreement efforts concerning areas of common transboundary air pollution, European Union efforts in many environmental arenas, and those of the Nordic states.

Other Factors   In analyzing application of new analyses and management in the Black Sea case, other factors make for a balanced, if not much more cautious, conclusion. Despite their forced union under the Soviet model, the riparians differ in many ways. When Turkey is added to this set, the cultural, religious, political, and historical differences among the nations become even greater. Systems of government, even within the set of former Soviet nations, have differed significantly. The history of shared cooperation and goals in the region and among the riparian states specifically is rather limited.

Conclusion

Sound environmental management of the Black Sea remains an immense challenge. It was so under prior regimes, and there are many reasons to be modest in expectations about opportunities for major shifts under the current regime. Governments still claim that they cannot afford to be environmentally sensitive when their economic conditions are miserable; the governments themselves are challenged by movements that are contrary to new attitudes toward inclusion and cooperation; the nations that are attempting to work together have significantly different political and cultural traditions; flows of information are impeded in several ways; the organizations that actually must manage are fragile; and meaningful public participation in transparent decision making lacks a historical base and a contemporary set of dispute resolution forums.

Nonetheless, efforts like those of the BSEP represent a potentially important change. The BSEP incorporates, at least at a rhetorical level, elements of a new understanding of transboundary interaction. It institutionalizes procedures that can be the core of productive linkages among the participating entities, the type of "indefinite iteration" essential to international cooperation. Careful nurturing of developing approaches, in the BSEP proper and in other initiatives in the region, involving new actors, new methodologies, and participatory values, remains necessary. In parallel, contributions from actors outside of the region can assist management. Considerable technical assistance is available through IGOs, NGOs, and universities worldwide. This expertise can be drawn upon to work with scientists and lawyers in the former Soviet Union and Turkey, emphasizing among other inputs public participation and openness to local knowledge. Many specialists would be pleased to apply their skills in an area as ecologically and technically challenging, intriguing, and beautiful as the Black Sea. Furthermore, successful international environmental regimes elsewhere can offer perspectives to the institutions of the region (chapter 7; Milich and Varady 1998).

One can envision a Black Sea whose environmental potential is understood in communitarian ways, whose resources are described as sustainable, whose managers solicit and welcome local information, perspectives, and data and that of cooperating neighbors. From the study of cases like the Black Sea—from the study factors linked to the functioning of regimes and the contexts in which those factors operate—we may be able to generalize to other areas where hostilities have been the tradition, where cultural and political barriers to communication have represented significant constraints, and where the environmental challenge is daunting.

Notes

1. The chapter is based in part on review of BSEP materials, including some which are remarkably self-critical and candid; interviews of participants and observers of the BSEP; focus group activities in Tbilisi and Sukhumi; and visits to the Programme Coordination Unit of the BSEP in Istanbul and activity centers of the program, including the Centers for Biodiversity and the Black Sea Ecology and Fishery Institute at Batumi.

2. Early reports of pollution by heavy metals and pesticides are countered by the *Black Sea Transboundary Diagnostic Analysis*, which concludes that "the concentration of . . . pesticides and PCBs . . . was found to be rather low in most cases . . . [and] it is quite apparent that the Black Sea is not generally polluted by heavy metals" (Global Environmental Facility 1997, 74).

3. At the Plymouth Environmental Research Center in the United Kingdom, Mee was recently awarded a Pew fellowship to address marine degradation in the Black Sea. He will now be involved in development and dissemination of a public version of the 1996 Black Sea Strategic Action Plan and organization of regional workshops for environmental educators, and he will examine the societal root causes of Black Sea degradation and explore possible solutions based on changes in values and attitudes.

4. Specifically, Article 67 of the *Strategic Action Plan* states that "by 1998, all Black Sea coastal states will adopt criteria for environmental impact assessments and environmental audits that will be compulsory for all public and private projects. The coastal states will cooperate to harmonize these criteria by 1999 and where possible, to introduce strategic environmental assessments" (1996, http://www.blackseaweb.net/action/content.html)

5. Reportedly the Programme was also evaluated after its three-year introductory period through the UNEP under Ellen Hey.

6. Recent surveys by the United States Information Agency found that majorities in the Russian Federation (65%), Ukraine and Kazakhstan said that they favored protecting the environment even if that meant slower economic growth for them.

7. Annual catch values "declined by at least a million between the early 1980's and mid 1990's" (Global Environmental Facility 1997, 21) and by "at least $300 million from the mid 1980s to early 1990s" (109).

8. The *Black Sea Transboundary Diagnostic Analysis* states that among the remedial actions required at the national and regional levels are "to establish ... legal basis for environmental NGOs participation in policy-making, implementation and assessment; ... to adopt ... legislation providing for the possibility to submit a law suit against a State official or State organ; to adopt ... rules obliging State officials to meet with the public on their request and to answer questions on environment ... to ensure ... open access to judicial organs, also in transboundary context" (Global Environment Facility 1997, 103). Each of these remedial actions was to have been completed by 1997.

9. The *Strategic Action Plan* (1996) concludes that "each Black Sea state shall endeavor to adopt and implement, in accordance with its own legal system, by 1999, the laws and mechanisms required for regulating discharges from point sources ... [and] will also endeavor to adopt and implement, in accordance with its own legal system, efficient enforcement mechanisms by 1999," III(A)(35)(c and d).

References

Ascherson, Neal. 1995. *Black Sea*. 1st American ed. New York: Hill & Wang.

Ascherson, Neal. n.d. "Can a Study Cruise and a Noble Scrap of Paper Save the Black Sea? Accessed 7 February 1998. ⟨http://www.hydrowire.org/internat/internat10.03a.html⟩.

Barcelona Resolution on the Environment and Sustainable Development in the Mediterranean Basin. 1995. Barcelona, Spain, June 10.

Belsky, Martin H. 1985. "Management of Large Marine Ecosystems: Developing a New Rule of Customary International Law." *San Diego Law Review* 22, no. 4 (July/August): 733–779.

Benvenisti, Eyal. 1996. "Collective Action in the Utilization of Shared Freshwater: The Challenges of International Water Resources Law." *American Journal of International Law* 90, no. 3 (July): 384–415.

Biermann, Frank. 1998. "Land in Sight for Marine Environmentalists? A Review of the UN Convention on the Law of the Sea and the Washington Programme of Action." *Revue de droit international de sciences diplomatiques et politiques* 1 (January–April): 35–64.

Brunnee, Jutta, and Stephen J. Toope. 1997. "Environmental Security and Freshwater Resources: Ecosystem Regime Building." *American Journal of International Law* 91: 26–59.

Caldwell, Lynton Keith. 1990. *International Environmental Policy: Emergence and Dimensions.* 2d ed. Durham, N.C.: Duke University Press.

Davies, Peter G. G., and Catherine Redgwell. 1997. "The International Regulation of Straddling Fish Stocks." *British Year Book of International Law 1996,* Vol. 67, pp. 199–274. Oxford: Clarendon Press.

Declaration on Euro-Mediterranean Cooperation on the Environment in the Mediterranean Basin. 1992. Cairo, Egypt, April 30.

Dellapenna, J. W. 1989. "Water in the Jordan Valley: The Potential Limits of Law." *5 Palestine YBIL,* 15–47.

DiMento, Joseph F. 1998. "Environment Focused Practical Peacemaking: Theory, Practice, Promise?" Mimeographed. University of California, Irvine.

DiMento, Joseph F., and Pamela M. Doughman. 1998. "Soft Teeth in the Back of the Mouth: The NAFTA Environmental Side Agreement Implemented." *Georgetown International Environmental Law Review* 10, no. 3 (Spring): 651–752.

Fuentes, Ximena. 1997. "The Criteria for the Equitable Utilization of International Rivers." In Ian Brownlie and James Crawford (eds.), *The British Yearbook of International Law 1996,* Vol. 67, pp. 337–412. Oxford: Clarendon Press.

Global Environment Facility, Black Sea Environmental Programme Coordination Unit. 1997. *Black Sea Transboundary Diagnostic Analysis.* United Nations Development Programme, New York.

GlobaLearn. 1996. "The Black Sea." Available: ⟨http://www.globalearn.org⟩.

Haas, Peter M. 1990. *Saving the Mediterranean: The Politics of International Environmental Cooperation.* New York: Columbia University Press.

Kliot, Nurit, and Deborah Shmueli. 1997. *Institutional Frameworks for the Management of Transboundary Water Resources.* Vol. 2. Edward & Anna Mitchell Family Foundation Water Resources Management Laboratory. Haifa, Israel: Technion, Israel Institute of Technology.

Milich, Lenard, and Robert G. Varady. 1998. "Managing Transboundary Resources: Lessons from River-Basin Accords." *Environment* 40, no. 19 (October): 10–15, 35–41.

Nollkaemper, A. 1996. "The Contribution of the International Law Commission to International Water Law: Does It Reverse the Flight from Substance?" In *Netherlands Yearbook of International Law,* Vol. 27, pp. 39–73.

Odessa Ministerial Declaration on the Protection of the Black Sea Environment. Odessa, April, 1993.

"opSEA: UNOPS Serving the Black Sea." 1998. Available: ⟨http://www.unops.org/finance/wbsea.html⟩.

Pergamon, Metropolitan J. 1997. "Religion, Science and the Environment, Symposium II: The Black Sea Crisis." September.

Sampson, Martin W. III. 1995. "Black Sea Environmental Cooperation: States and 'The Most Seriously Degraded Regional Sea.'" *Bogazici Journal: Review of Social, Economic and Administrative Studies* 9: 51–74.

Sampson, Martin W. III. 1996. "Environmental Aspects of Migration in the Black Sea Region." Paper presented at the Conference on Migration and Security, Istanbul, September 1996.

"State of the Environment Georgia." 1996. Geoinformation Center, Tbilisi, Georgia. Available: ⟨eis@ginfo,kheta.ge⟩.

Strategic Action Plan for the Rehabilitation and Protection of the Black Sea. 1996. Istanbul, Turkey, October 31. ⟨http://www.blackseaweb.net/action/content.html⟩.

United Nations Development Programme (UNDP), United Nations Environmental Programme, The World Bank, Phare, and Tacis. 1997. *Black Sea Environmental Programme: 1997 Annual Report.* UNDP, New York.

Vinogradov, Sergei. 1996. "Transboundary Water Resources in the Former Soviet Union: Between Conflict and Cooperation." *Natural Resources Journal* 36, no. 2, part 1 (Spring): 393–415.

10

Water as a Boundary: National Parks, Rivers, and the Politics of Demarcation in Chimanimani, Zimbabwe

David McDermott Hughes

Water exists as both a volume and a line.[1] As volumes, acre-feet or liters of water hold value for irrigation, hydropower, and similar uses. So valuable, in fact, is volumetric water that corporations, governments, and individuals fight over the terms of its acquisition, sale, and transport. With engineers' disregard for landscape, they transfer water far and wide and always count how much. The bulk of the writing on water politics—summarized in chapter 2—has concentrated on issues of commoditized, movable volumes abstracted from their home terrain. Yet water is also and perhaps more pervasively a line around and through places. Where water moves as river or stream, or even where it once moved and created a track, it marks a line across the landscape. Like a fence, this kind of linear water may serve to delimit a given territory or resource; it lies at the root of land ownership and land disputes. When measured in meters, rather than in cubic meters, water animates the politics of territorial control.

Nowhere is it more likely to do so than in the mountains. In uplands, such as those of Zimbabwe's Chimanimani (figure 10.1), water is abundant per capita, watercourses are dense and stable, and surveying costs are high. On the plains, where water economists have formed many of their assumptions, the conditions are reversed: Water is relatively scarce per capita, watercourses are fewer and are less stable, and surveying is cheap and easy. Let us take these attributes one at a time. First, fewer people live in the mountains, but mountains often absorb the bulk of rainfall. The less well-watered plains, by contrast, contain most cities. On a per-person basis, the availability of water declines with lower elevation. Second, mountains' rugged topography distributes that precipitation

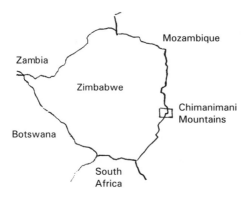

Figure 10.1
Chimanimani, Zimbabwe

into a dense network of linear water: streams, rivulets, and tributaries. En route to the plains, however, water aggregates into a few major rivers. Third, broken ground canalizes linear water. Concentrated in a ravine, water meanders less than it does on a deltaic floodplain. Fourth and finally, technical surveys in the mountains confront tremendous obstacles. Craggy, forested terrain imposes a unique misery on anyone who would walk a compass bearing or measure a straight, horizontal distance (unless he or she possesses the latest lasers). For this reason, any practical geographer will find natural features, particularly fixed, deep-cut watercourses, to be the easiest, least painful marker of boundaries and other lines. On the plains, obviously, surveys can mark and measure straight lines with minimal hardship. In short—but without any firm determinism—mountain environments encourage linear uses of water, whereas flatlands dispose one to use the same liquid as a volume.

These distinct roles for water, in turn, shape regional history. Politics on the plains have often centered on that quintessentially volumetric practice, irrigation. As chapter 3 explains, the Mexican and U.S. governments have made social policy for the arid Imperial and Mexicali Valleys by allocating, conserving, and moving acre-feet of water derived from the Colorado River. They have variously constructed water as a product, as security, as the means of social redistribution, and as a commodity. By contrast, the regional history of alpine landscapes may turn on the bounding properties of linear water. This chapter examines the politics of linear water in an area called Vhimba at the southern end of the Chimanimani Mountains in Zimbabwe. In the well-watered Chimanimanis, the powerful and weak have struggled not over the allocation of water but over the allocation of land. In the 1890s, white settlers appropriated most of the upland. In the 1960s and 1970s, the white-dominated Rhodesian government took more land from African smallholder farmers in the name of conservation. A war of liberation and national independence in 1980 failed to redistribute either private estates or any part of the Chimanimani National Park and associated botanical reserves. At the southern end of that park, where the Rusitu River knots with its Haroni and Chisengu tributaries (figure 10.2), smallholders are now fighting some of Zimbabwe's most bitter land disputes. Smallholders claim alienated land on the basis of linear water. The government and a private landowner, on the other hand, justify their authority over the same parcels with reference to artificial, surveyed lines. The rest of this paper explains the origins of Vhimba's two principles of demarcation—linear water versus survey lines—and details the conflicts that have ensued therefrom.

How Rivers Came to Be Boundaries

Various sources indicate the importance of riverine boundaries in Chimanimani before its colonization by whites in 1893. Between the 1830s and white occupation, two kingdoms partitioned much of what is now Zimbabwe. The Ndebele and Gaza Nguni states, both derived from northward Zulu conquests, established the Mutirikwi River as the boundary between their zones of raiding (Latham 1970, 26). Under these kingdoms, dynasties ruled by chiefs occupied hilly outcrops and held authority over neighboring hinterlands. David Beach describes the political geography of Shona-speaking south-central Zimbabwe in the late nineteenth century as follows: "In the lowveld, territorial boundaries were rather vague and enclosed large areas in which the people often moved long distances in order to find water, grazing or game. In the mountains or on the plateau, territories tended to be smaller, and demarcated by definite borders along streams or ridges" (Beach 1986, 49). Ridges, since they divide watersheds, are an oblique form of linear water;

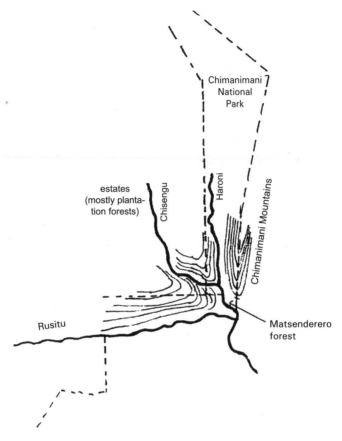

Figure 10.2
Chimanimani National Park

so, at least in Beach's area, Africans marked territory aquatically. To the east, Vhimba lay on the threshold of the Gaza Nguni state, under the authority of the nominally independent Chief Ngorima.[2] Territorial notions there were probably more latent than active, for Nguni raids confined Ngorima's people to caves and other hideouts in the Chimanimani Mountains.[3] As Britain imposed colonial peace in the 1890s,[4] Chief Ngorima organized his authority within riparian boundaries: He established what Thongchai (1994) describes as a "geo-body," a territory identified with a polity and lying within fixed, clear lines. White colonization both necessitated this kind of formalization and constrained Ngorima in the amount of territory he could claim. Beginning in 1893, settlers from what is now South Africa took Chimanimani's best soil. Largely Afrikaaners, they sought fertile land for homesteading and were happy to evict Africans or transform them into tenant farmers. Having struck out in advance of government supervision, these frontiersmen almost appropriated the totality of Ngorima's geo-body. In response, the colonial administration established in 1900 a reservation[5] for the Ngorima "tribe" on the fringes of their previous zone: the northern bank of the Rusitu River. On the whole, a hundred or so whites retained title to roughly half of the Chimanimani area. The patrimony left to African smallholders consisted of low-lying river valleys rife with cattle diseases and malaria (Rennie 1973, 179–180). If Ngorima's remaining geo-body was small, it was unusually rich in linear water.

In the course of the next half century, Ngorima organized his administration in a fashion that made optimum use of water's bounding properties. The colonial administration recognized him and, under the principle of indirect rule, paid Ngorima a salary for his services as a local ruler. He bore responsibility for collecting taxes and marshaling his subjects for public works. His court ruled on minor disputes of marriage and property that did not demand a magistrate's involvement. Ngorima and his advisors also regulated the entry of migrants into the reservation. Given whites' expropriation of land nearby, this integrative function became all-important and placed more and more emphasis on river boundaries inside the reservation. In theory, Ngorima should have allocated parcels of farmland within the reservation to all suitable newcomers, but the number of newcomers and the amount of land available exceeded the practical capacity of one land allocator. Therefore, by the 1940s, Ngorima (now the grandson of the titleholder from the 1890s) had partitioned the reservation (and some adjoining private land he considered his) into subjurisdictions under headmen. The chief chose to mark these sections with the best means available to him on the landscape: streams. As drawn in a 1976 map by district government (figure 10.3), watercourses set off four headmanships of roughly equal geographical area and population. The map also indicates a further refinement: the larger watercourses bound more important political units such that the proper rivers—the Rusitu, Haroni, and Nyahode—form

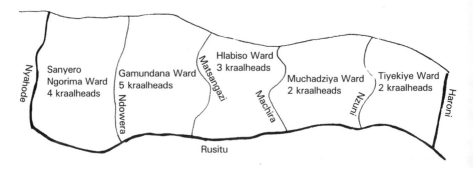

Figure 10.3
"Ngorima TTL" (Reservation) 1976
Source: Offices of Chimanimani District Administrator, Ngoirma PER file.

three of the four sides of Ngorima's domain (although the legal western boundary actually followed a straight line rather than the Nyahode). At the other end of the political hierarchy, streams too small to appear on the map set off kraalheads under each headman. Thus Ngorima graded, divided, and subdivided the smallholder population along a grid of linear water.

In the course of the second half of the twentieth century, headmen (and kraalheads) have used this grid to their advantage. Widespread evictions from private estates provided them with a mechanism for doing so.[6] At roughly midcentury, the descendants of the white settlers sold their bankrupt farms to large timber corporations. Long-standing African tenants had to make way for intensive silviculture. They had to find new lands in the Ngorima reservation, and headmen helped insert them in the grid of linear water. Ever since then, each headman has told migrants where to live and where to farm (or at least limited the number of possible parcels) within his grid space.[7] They have done so, moreover, in the teeth of government legislation. The Native Land Husbandry Act of 1954 intended to replace headmen's allocation of land with technical land use planning. This reform and similar efforts in the 1980s failed in Vhimba largely because headmen would not relinquish their territorial powers.[8] Indeed, headmen have often sought to expand their powers of land allocation by grabbing territory from each other. In such cases, headmen disagree over indistinct segments of Ngorima's otherwise re-

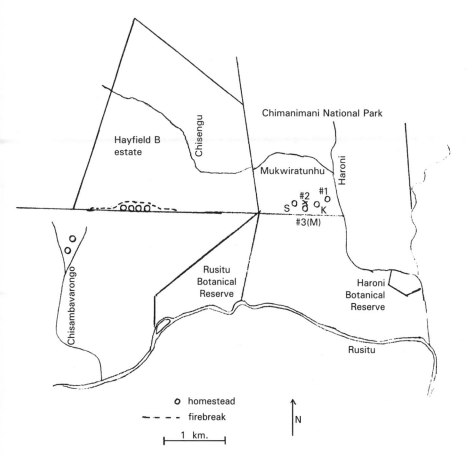

Figure 10.4
Vhimba

markably clear riverine grid. Dominance, then, turns on the location of the chiefdoms' internal boundaries, and those boundaries are and have been linear water.[9]

During the time of my fieldwork in Vhimba in 1995–1997, Headmen Chikware and Muitire were engaged in just such a dispute. Because Chikware and Muitire were kraalheads—too insignificant to have appeared on the 1976 map—their mutual border followed quite a small stream, the Chisamabavarongo (figure 10.4). Although at its downstream end, the Chisambavarongo runs year round, along the escarpment of the Rusitu Valley it nearly peters out.

The turf battle between Chikware and Muitire arose because, close to its headwaters, the Chisambavarongo, as the longest tributary of the Rusitu, becomes indistinguishable from another long tributary. Precise measurement was nearly impossible, so each headman identified as the true Chisambavarongo the watercourse that bounded a larger territory for himself. That is, Muitire, as the western of the two headmen, claimed the eastern watercourse, whereas Chikware claimed the western one. In 1991 and 1992, Chikware put his assertion to the test by allocating land in the triangle between the two streams to two migrant households. This fait accompli seemed to establish the border as the eastern stream. Yet five years later, Muitire, Chikware, and their partisans still argued about linear water. They were waiting for Chief Ngorima to adjudicate the dispute, to "fix the boundary" (*kugadzira muganhu*).

If riverine demarcation thus set headmen against each other, it also frequently united Vhimba's people against the state. Fundamentally, the geo-body claimed by Chief Ngorima and his headmen extends northward beyond the official boundaries of the reserve. Hence, many headmen and kraalheads claim to have lost land north of the ridgeline that now delimits the Ngorima Native Reserve. In the worst case, a forestry estate took all of Headman Parara's land, thereby demoting him to plebeian status. "My country was taken by a farm," he said, "... I [then] lived in the area of [Kraalhead] Matwukira.... I am ruled. [When told], 'do it,' I do it right away."[10] Against this kind of land-grabbing, ex-headmen, reigning headmen, aspirant headmen, and commoners have insisted upon their place and manner of demarcation. To their mind, linear water—especially when sanctioned by the chief—should mark the divisions within Ngorima's domain and between that territory and other categories of land. Upon this much, Chikware and Muitire had certainly agreed. Rivers were customary and just, whereas artificial lines ran counter to Vhimba's sense of right. This moral approach brought the population of the reservation into conflict with the state at two levels: Vhimba residents resented the location of boundaries of alienated land, and they disputed the entire rationale of those surveyed lines.

How Straight Lines Came to Be Boundaries

Surveying arrived in Chimanimani hand in hand with settler farming. When, in 1892, the British South Africa Company authorized a "trek" of Afrikaans-speaking farmers, it delegated to the organizer of that trek, G. B. D. Moodie, the responsibility for allocating land. Unlike headmen, Moodie was to designate parcels according to precise stipulations that contained no reference to linear water. Indeed, the terms of white occupation of Chimanimani deliberately occluded the aquatic grid. To mark property lines, read Moodie's instructions: "beacons ... [are] to be placed on conspicuous ground ... and the straight line between the two beacons is to be considered the boundary."[11] "Conspicuous ground" meant an elevated area, presumably between two valleys that would have held water. Hence the boundaries connecting beacons would necessarily cross rivers and streams. The use of straight-line boundaries, of course, completely ruled out demarcation along *curvi*linear water. The "allotment of farms," then, would necessarily forego the bounding possibilities offered by Chimanimani's landscape and already known to its inhabitants. Fortunately, Moodie came from a surveying family (Burrows 1954, 109), and he had no difficulty "pegging" the estates as instructed. As his cadastral map of 1894 shows (figure 10.5), the settlers established rectangular farms cross-cutting and bisecting the Rusitu and every other watershed. Whereas headmen eventually laid down a wavy lattice of streams, white colonization established a grid much closer to the ideal type of parallel lines intersecting at right angles.

What land alienation achieved in Chimanimani more generally, the native reservation system brought to the Rusitu Valley and Vhimba specifically. Uncertain of the eventual location of the Portuguese border, homesteaders spared the Rusitu Valley, making it available for a reservation. In 1892, the British South Africa Company had cautioned Moodie, "you will be safe in treating the whole of the High Veldt as British South African territory, and the Mozambique the Low Veldt."[12] Hence, Moodie's allocation of farms left the Rusitu Valley and intersecting watercourses to the Portuguese. In 1897, the Anglo-Portuguese Delimitation Commission placed the boundary at the Rusitu River, leaving its northern bank inside British territory and free for native

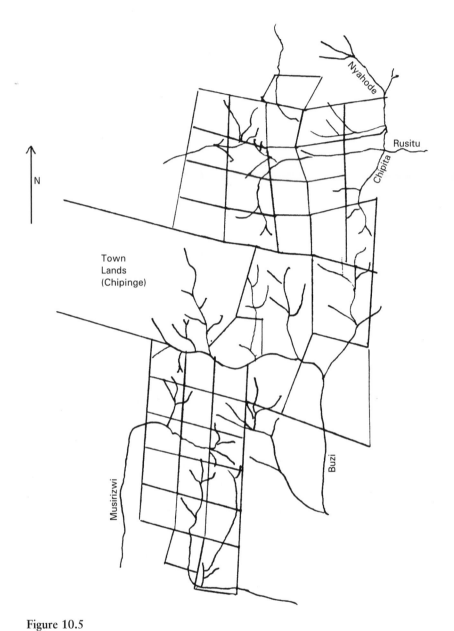

Figure 10.5
1894 cadastral map
Source: National Archives of Zimbabwe. File L2/2/95/25, "Map showing approximate position of farms taken up in Melsetter, Gazaland," January 26, 1894.

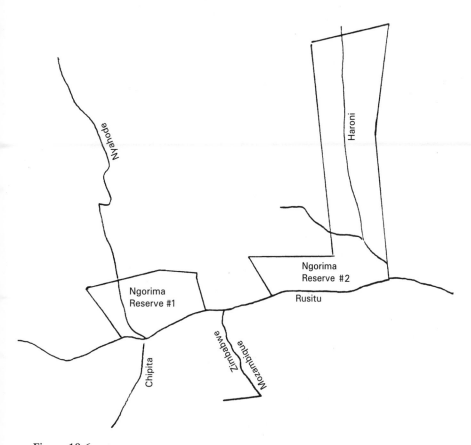

Figure 10.6
Ngorima Reservation
Source: National Archives of Zimbabwe. File N3/24/20, approximately 1914.

occupation. This set of circumstances resulted in a reservation rich in linear water. Again, though, the administration took no advantage of these bounding possibilities. The administration drew an elbow-shaped zone, delimited on five of its six sides by straight lines (figure 10.6).[13] In part this result followed from Moodie's method of allocating white farms: alienation by survey logically created surveyed lines for the African enclaves that remained. As Surveyor General R. W. Sleigh later reflected—without a hint of irony—"Surveyors and their discipline ... have served Rhodesia well in the past and continue to do with notable success as an important pillar in the creation of prosperity for all" (Sleigh 1976, 4).

In the subsequent course of the colonial period, Chimanimani's land owners made good on their straight lines by evicting African "transgressors." Moodie and his co-colonists allowed Africans to remain on the Chimanimani plateau as farm laborers and tenants, cultivating the unused fringes of these underutilized estates. Indeed, these economically marginal farms needed a supply of cheap resident labor. Yet even that subsidy could not prevent the capital-starved estates from going bust as mid-century approached. In the 1950s, as mentioned above, large corporations bought out the trekkers' descendants to establish plantation forestry. Thenceforth, pines and other exotic trees blanketed the plateau, forming a neat wall along the northern line of Ngorima Reserve (figure 10.2). The new plantation owners tossed Africans over the wall. Silviculture required less unskilled labor and more land than had white agriculture, so the black tenant farmers and farm laborers would have to go. As one evictee, now living in Vhimba, recalled: "We had a big place. It was taken from us. We are left with a very small place."[14] Many—whether actual evictees or sympathizers, we do not know—resisted the plantations and burned the forests. In 1962, "a handful of fire-raisers defied the organised and armed forces of the law and threatened Rhodesia's timber industry" (Sinclair 1971, 173). Military patrols then guarded against *moto mukuru*, "the big fire" (Sinclair 1971, 173; Godwin 1996, 119–120). Thus, the Rhodesian state made its point: North of the escarpment, Africans had no rights to land or trees.

Winning the whites-only election of 1965, the uncompromising Rhodesian Front Party drove this point home. The state, represented by the Department of National Parks and Wildlife Management, took *more* land from Chief Ngorima and his people. Although now justified in the name of conservation preservation, these acts of land grabbing opened new vistas for white tourists' recreation and for the (white-owned) hospitality industry's profit making. Once again, land alienation followed the survey. Having opened in 1953, the Chimanimani National Park expanded in 1965. Extending southward, it engulfed the Haroni Valley, the adjoining slope of the Chimanimani Mountains, and a patch of lowland moist forest in Vhimba (figures 10.2 and 10.4). That woodland, the Matsenderero Forest, lay across the Haroni River from the more densely populated parts of Vhimba, so the park's new southern boundary followed the rivercourse for a stretch. West of the forest, however, the boundary diverted from the Haroni and climbed a steep slope to join a corner beacon of the Hayfield B estate. Artificial lines also bounded two botanical reserves declared in 1973 (figure 10.4). The cadastral descriptions of the Haroni and Rusitu Botanical Reserves relied almost entirely upon compass bearings. The former followed the Haroni River for two of four sides but otherwise struck out along survey lines. Also quadrangular, the Rusitu Botanical Reserve anchored itself to the Rusitu River, but between the river and the aforementioned corner beacon of Hayfield B, cut three straight lines up the forested, steep slopes of the escarpment. Fortunately for people in Vhimba, the Department of National Parks enforced none of these boundaries in the 1960s and 1970s. Guerrillas fighting to liberate Zimbabwe from white rule occupied the area and reduced the demarcations to mere "paper parks." The holiday lasted until 1980, when a black government took power and began to evict "squatters" from the parks.

By this time, Vhimba residents had formed a clear notion of just and unjust boundaries. Unjust boundaries were those through which white settlers, timber companies, and state-backed conservationists alienated land. Such lines of eviction also happened to be straight, artificial, and surveyed. The pattern was general in Rhodesia, such that, at Zimbabwe's independence in 1980, native reservations held only 42% of the land area of the land area of Zimbabwe and only 27% of its most arable land categories.[15] Not only were these reserves bounded by lines, but many of them organized African farming along a grid. The Native Land Husbandry Act of 1954 and subsequent efforts at land use planning had allocated arable and grazing plots arranged in rows along access roads. In many parts of Zimbabwe, *maraini*, or "the lines," constituted nearly as strong a point of resistance to Rhodesian rule as land alienation itself (Moore 1995, chap. 5). Vhimba residents thus associated straight-line planning with bureaucratic intervention and territorial theft. In the 1990s, they would express their opposition to state-backed land grabbing by advancing their own idea of legitimate, natural, aquatic boundaries.

Turf Battles and Contested Borders

Vhimba residents had every reason to believe that the post-independence black government of Zimbabwe would restore land taken from Ngorima.

ZANU (the Zimbabwe African National Union) had declared its intention to redistribute land. Vhimba residents helped the guerrillas infiltrate
Zimbabwe from Mozambique and were assured that, from then on,
"you may live where you want to."[16] Presumably, this policy would have
allowed headmen to allocate land within the alienated spaces and up
to the riverine boundaries the headmen claimed. Once in power, however, the new government did little to return the Chimanimani plateau to
headmen or to any other of Ngorima's people. Slightly to the east, in
the Nyahode Valley, the state did allow smallholder farmers to settle
on farms abandoned by war-weary whites. This area was too far from
Vhimba, and the strict legal conditions for resettlement prevented all but
a handful (if that) of Vhimba's families from acquiring land there.[17] In
the meantime, the population of Vhimba grew through natural increase
and in-migration. Squeezed into the reserve, neighbors encroached upon
each other's fallows, and headmen were forced to allocate increasingly
marginal and undesirable land to newcomers. In 1995, Vhimba's farmers were strung along steep slopes and precarious ridges, within sight of
empty public and private land. Palpably affected by the injustice, one
farmer asked me, "In what way have those without a place to live been
liberated?" ("*Vanhu vasina pokugara vanosunungaka chii?*") Clearly,
headmen, the government, and other land managers would clash, and
clash along the predictable fault lines of contested boundaries. These
demarcations, drawn in places where straight lines are virtually unwork-
able, trace the Hayfield B estate, Chimanimani National Park, and the
Rusitu Botanical Reserve.

The Hayfield B estate constituted the first and largest parcel of alienated land whose boundaries came under debate. Here, land use and
topography threw into stark relief the contradiction between linear
water and lines pure and simple. Grabbed close to the turn of the century (although then part of the Tarka estate) along with the rest of the
plateau, Hayfield B extended from the ridgeline of the Rusitu watershed
northward across the Chisengu River and to the opposite upland (figures 10.4 and 10.6). Unlike the estates to the west and north of it,
Hayfield B had escaped conversion to plantation forests. In fact, the current owners of the estate wished to spare it from planting of any kind.
Concerned with preserving biodiversity, they planned to manage the

Chisengu Valley as a private butterfly sanctuary. This orientation cast the tree-lined Chisengu River as the property's heartland. Vhimba residents also recognized the importance of this watercourse, but as the edge, not the core, of a territory. It marked the northern limit of Headman Chikware's area, a mounded expanse stretching from the Rusitu to the Chisengu. Indeed, influenced by guerrilla "comrades," Chikware and his people had decided to farm the southern bank of the Chisengu. Between 1975 and 1981, a handful of households had come over the estate border, and two homesteads had cultivated a wide swath of the Chisengu Valley.[18] Government officers expelled these "squatters" in the early 1980s. Except for one household mysteriously spared removal, Vhimba residents remained south of the surveyed line for a decade.

In the early 1990s, however, land pressure forced Chikware to try to move the border northward again and to compromise with the estate's owner. Chikware settled three families in an amphitheater-shaped area around the remaining squatter. All four households thus appeared to lie north of the surveyed line. Of course, no one knew for sure where this artificial boundary lay. If the smallholders had encroached, they had done so by less than 100 meters, and they still lay south of the much more visible ridgeline dividing the Rusitu and Chisengu watersheds. Unexpectedly, the estate owner also cared much more about this latter boundary than he did about the official demarcation. As long as smallholders stayed out the Chisengu watershed, he felt, erosion would not threaten the butterfly sanctuary. To protect it from another danger, he cleared a fireguard along the ridgeline. Both parties to the dispute thus converged on a hydrological definition of their border. The estate owner relinquished the straight-line boundary that marked his legal tenure. For Chikware, though, the sacrifice was greater. He abandoned the notion of a territorial unit constituted by upland and bounded by rivers and, in its place, acknowledged the exact inverse: Hayfield B as a territorial unit constituted by the river valley and bounded by the headwaters. What had been the core of Chikware's area became its edge, and what had been the edge of Chikware's area became Hayfield B's agreed-upon core. In effect, the estate owner had waded into the topographical, water-based boundary making of Vhimba and emerged triumphant. Confident in this victory, he responded in two words to the petition from a former evictee from the Chisengu Valley: "No way!"[19] Butterflies claimed precedence over farmers.

Also related to conservation, the turf battle involving Chimanimani National Park sprang from the same basic predicament as Hayfield B. Extended southward in 1965, the park took from the Vhimba community the whole of the lower Chisengu Valley downstream from Hayfield B (Hughes 1996). This area, called Mukwiratunhu (figure 10.4), contained neither the moist lowland forest nor the stunning mountain outcrops for which the Chimanimani Park is known. Rather, it seems that surveying conventions alone caused the park to engulf Mukwiratunhu and its hapless inhabitants. The park's new straight-line boundary followed the same ridge bounding Hayfield B, but only approximately. The demarcation, therefore, reproduced all the uncertainties inherent in artificial, invisible lines. To make matters worse, the precise wording of the park's technical description treated Hayfield B's boundary as a certainty and therefore magnified the scope for error. The demarcation between the park and communal land traces the "eastern prolongation of the southern boundary of Hayfield B [down to the Haroni Valley]" (Rhodesia Act 5 of 1979, Land Tenure (Repeal), 116). Without an anchor beacon on the Haroni, the angle of that survey line is almost unverifiable and, by simple geometry, its position grows less and less knowable with each eastward meter.

Nonetheless, this uncertainty did not provoke conflict until Zimbabwe's independence. Prior to the 1980s, Rhodesia's Department of National Parks and Wild Life Management had been content to allow "squatters" in the park so long as they stayed south of the Chisengu River, that is, in Mukwiratunhu. Although technically inside the park, Mukwiratunhu was completely inaccessible by vehicle, and, by the 1970s, guerrillas discouraged government employees from serving eviction notices there. Ironically, the black government took up this unpleasant task. Ejecting three families between 1981 and 1987, the post-independence Department of National Parks cleared the southern bank of the Chisengu. From then on, Vhimba residents accused the ZANU government itself of having moved the border from the Chisengu to the more southerly line. In the words of one critic in Vhimba, "The border was taken by the Comrades" ("*muganhu wakatorwa nemakamba*").

Indeed, the government had nullified a clear aquatic boundary and re-placed it with only a vague "forbidden zone," the fuzzy vicinity of an indiscernable straight line.

Into that zone, two headmen sent three distinct waves of land alloca-tees. In 1991 and 1992, Headman Tiyekiye settled four households in Mukwiratunhu. He chose to settle migrants in this area because "it was inhabited before. Mafuta died there. Maparara died there too."[20] Ironically, Tiyekiye's allocations provoked opposition from a second headman, Muhanyi, even before the Department of National Parks in-tervened. Muhanyi served under Tiyekiye. As a kraalhead, he had not appeared on the 1976 map of traditional jurisdictions. Nonetheless, Tiyekiye had delegated to him informally the task of managing in-migration to Mukwiratunhu. Now, seeing this responsibility reclaimed from him, Muhanyi quickly settled, by his own authority, four addi-tional families in Mukwiratunhu. To hammer the point home, he posted a sign in Mukwiratunhu advising, "Let us try to do the wishes of the owner of this place, David Muhanyi. Live well with others in this place. Care for your livestock."[21] Tiyekiye backed down; he ceased to allocate land in Mukwiratunhu and allowed Muhanyi to act as headman toward the four original families. The Department of National Parks, as well, seemed to acquiesce to the headman's recolonization of old lands. Un-disturbed by park officers, many of the "squatters" did not even know they farmed along or over a boundary.

In 1993, however, the Department of National Parks again turned its attention to peripheral areas. The fact that many of the land allocatees were Mozambican refugees—having fled that country's drought and civil war[22]—provided a legal justification for summary eviction without re-course. Over the next two years, a campaign of harassment by enforce-ment officers contributed to the departure of six of the eight offending families. The remaining two dug in their heels, insisting (in the words of one head of household), "Mukwiratunhu is our place. It is not a park" (*"Mukwiratunhu nzvimbo yedu haisa parka"*). In fact, officers permitted these two households to stay, not because the department acknowledged their rights, but because the families inhabited a gray area surrounding the straight survey line. No one could be sure whether the suspects lived north of, south of, or smack on the boundary.

To Muhanyi, this tolerance suggested a new strategy: to move the boundary incrementally northward. When the next small batch of settlers arrived, he did not thrust them deep into Mukwiratunhu towards the Chisengu River, as before. Instead, he used them to make a compromise. He placed them slightly within the park, as markers of a conciliatory, artificial boundary. In late 1996, he placed Chirwa[23] at point 1 (on figure 10.4). Together with the two squatters of longer standing, Chirwa formed a line roughly east to west, parallel to but 200 meters or so north of the official survey line. When Mavi arrived in early 1997, Muhanyi sought to complete that line by siting him at point 2, directly between the two older residents. As one nearby farmer recalled, the headman "wanted to plow a boundary" so that the settlers "made a single line."[24] Ultimately, the plan was not entirely successful because M. did not want to leave his wife on that northern fringe where she would be exposed to wild animals and other dangers of the bush.[25] Hence, M. negotiated with Headman Muhanyi for an area just slightly to the south, at point 3. Thus Muhanyi created a zig-zag rather than a straight line. Still, he had abandoned linear water as a benchmark for territorial control and, on these unfamiliar terms, he had won a small a victory.

Rusitu Botanical Reserve

In the final boundary dispute, Vhimba people have refused to compromise on their hydrographic demarcation. Surprisingly, they have escaped eviction completely. In part, linear water has proved compelling because the Rusitu Botanical Reserve relies upon the most impractical survey ever conducted in Vhimba. Created in 1973, the reserve was meant to protect a patch of high-canopy, moist forest, one of only a handful of such patches in Zimbabwe. Yet the Department of National Parks chose not to designate the odd-shaped tree-covered zone itself. Instead, it instructed surveyors to map a territory between the Rusitu River and three artificial lines (figures 10.2 and 10.4). And these lines are artificial in the extreme. Whereas similar lines mentioned above at least approximate horizontal ridges, these lines ascend a slope of 35% (a rise of 520 meters over a horizontal distance of 1400 meters) on average and then plunge back down, through thick forest and over cliffs. How could one ever mark straight lines on a landscape such as this? In fact, in the 1970s, no one ever verified the table survey on the ground. Quite apart from topographical inconveniences, guerrillas kept the surveyors out of the Rusitu Valley. At most, the Department of National Parks indicated the boundaries hastily and imprecisely with ephemeral cut lines on trees. Even after independence, the department neglected this remote rainforest for more than ten years. Finally, as part of its 1993 crackdown, the department arranged to resurvey and mark the boundaries. In 1994, a surveyor more or less "ground-truthed" the technical description, allowing the department's enforcement officers to charge sixteen families with squatting. Yet the terrain had necessitated a rather large margin of error: distances measured in 1994 differed by as much as 55 meters from the technical description whereas compass bearings diverged by up to 1.5 degrees. The latter distortion, amplified over a distance of 2250 meters, led to a further discrepancy of 58 meters. Finally, the legal description relied upon an island in the Rusitu River whose shape had changed since 1973. These manifold ambiguities opened up a cadastral gray zone wide enough to swallow the field and homesteads of a number of alleged squatters. Straight lines served the purpose of marking the reserve's boundary very poorly indeed.

Despite—or perhaps because of—these technical failures, the demarcation generated a violent turf battle. Shortly after the resurvey, officers of the Department of National Parks began a campaign of harassing the families inside the surveyed boundary. They assaulted and arrested at least one farmer whose banana field appeared to cross the invisible line. This strategy did not work. In early 1995, the Department recognized that "the recent demarcation of boundaries ... has precipitated unprecedented hostility from Chief Ngorima and the local community."[26] Ngorima opposed the Department of National Parks on three levels. First, of course, the survey indicated that the state intended to take actual control of land. Second, he and Vhimba's headmen considered the forest in question to be sacred. Headman Chikware's ancestors are buried there, and the Headman still conducts annual ceremonies in their memory. How, Vhimba people asked, could the state take such a forest away from them? Responding to this outrage, Chief Ngorima's spokesperson used the occasion of a succession ceremony in 1995 to appeal directly to the Minister of Lands: "We ask you Comrade Minister [Kumbirai Kangai] for our important places (tribal monuments), which are ... [six sacred sites and] the forest of Nyakwaa in Vhimba.... We can care for these places so that our land stays good and is given blessings by the Creator."[27]

Third and finally, Ngorima, headmen, and Vhimba people universally opposed the botanical reserve because of its association with eviction. No one in Vhimba (or in government at that time) detected the weaknesses of the survey. They simply refused to accept the boundaries it had laid down. Headman Chikware continued to conduct rituals in the forest without seeking a permit from the Department of National Parks. He also refused to resettle any of the "squatting" families. Thus, the year 1995 passed without any resolution or compromise.

In 1996, however, a new set of actors from Vhimba decided to tackle the boundary question head on. A few years beforehand, a nongovernmental organization (NGO) had assisted some of Vhimba's prominent men and women to form the Vhimba Area Development Committee. With funds from the Australia Commission, the NGO had hoped to nudge Vhimba away from territorial struggle and toward some form of joint enterprise, such as eco-tourism, with the Department of National Parks. The committee had seemed interested. Yet now it shelved those ideas and threw itself into the fray over the Rusitu Botanical Reserve.[28] To the cause of Vhimba's land claimants, it contributed an unprecedented tactic: making a map. The committee argued that the Parks Department's 1994 survey had diverged radically from the cut lines of 1974. It asserted that those marks had followed linear water, a dry streambed to the west and a wet one to the east. The chair of the committee railed, "There is no boundary that is a pathway [i.e., artificial]. The boundary is the stream" ("*Hapana muganhu wenzira. Muganhu ngechimvura*"). Indeed, a contemporary member of the district government had the same recollection: that watercourses served as the first, approximate boundaries. They therefore resolved to "erect our original boundaries" through a map (minutes of meeting of Vhimba Area Development Committee, Vhimba, November 3, 1996).

Quite comprehensive, that document shows three sets of boundaries and sixteen households (figure 10.7). From the interior outward, it indi-

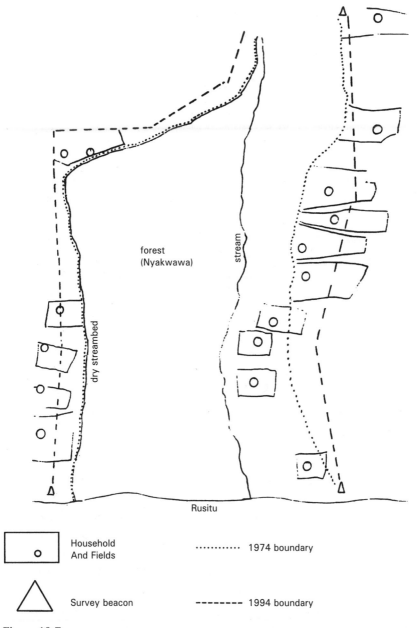

forest
(Nyakwawa)

stream

dry streambed

Rusitu

	Household And Fields	············ 1974 boundary
△	Survey beacon	- - - - - - - 1994 boundary

Figure 10.7
Rusita Botanical Reserve and Nyakwawa (sacred forest). Drawn by Elias
Nyamunda, Secretary, Vhimba Area Development Committee.

cates the edge of current forest, the 1974 limit of the botanical reserve (which coincides with the forest fringe on the west), and the 1994 surveyed limit of the reserve. Between the forest and the eastern 1974 boundary, the map places four families. Between the two sets of reserve boundaries, it shows the fields and homesteads of seven families. Finally, it displays a further five households resident outside the 1994 boundary but holding land inside the disputed, interborder strips. The map, then, demonstrates that Headman Chikware and his land allocatees have largely respected the original, marked lines of the National Park zone. Cartography exonerated the squatters. It also upheld linear water—represented as the 1974 boundaries—over straight lines. How could the state argue with such a clear demonstration? It could not: In late 1996, district government canceled the meeting at which the committee was to present its work to the Department of National Parks.[29] Neither boundaries nor farmers have moved since.

The turf battles described above are striking for two reasons, one expected and one unexpected. First, as anyone could have predicted, the owner of Hayfield B and the Department of National Parks clashed with Vhimba people over demarcatory logics. The opposition of straight lines and linear water was sharpest in the case of Hayfield B: The river lay at the center of the owner's claim and formed the boundary of the headman's claim, whereas the ridge line constituted the owner's boundary and the headman's territorial interior. One man's heartland was another man's hinterland and vice versa. Surprisingly, however, this headman and his peers found compromises and innovative strategies to mediate between straight and riverine lines. They devised an artificial means of marking the landscape but one that possessed all the ease and visibility of aquatic demarcation. Except in the case of the Rusitu Botanical Reserve—where the high number of "squatters" ruled out any compromise—headmen relinquished aquatic boundaries. Instead, they settled newcomers in a fashion that established territorial lines. In essence, land allocators used farmers' huts and fields as boundary beacons, a low-budget fence. In technical terms, this kind of boundary is not quite as good as linear water: Huts and fields are not permanent, nor can they be used to claim land that is too steep or rocky to inhabit or farm. Yet "residential bounding" certainly made territorial limits manifest in way that the woefully inadequate surveys had failed to do. Most important, this form of demarcation, like linear water, represented a technology that headmen and their subjects could master. Whereas surveying—the colonizer's science—had relied upon specialized equipment and training, Vhimba's methods of "barefoot" surveying were remarkably transparent.

This observation—a question of measurement—recalls the initial contrasts between volumetric and linear uses of water. Practices such as irrigation and bounding rely upon distinct means of measuring water. They also entail different objects of contestation. Irrigators use flow meters and depth gauges to determine the number of liters moving or stored. They then frequently assign a value to each liter and sell or tax the user accordingly. As chapter 3 shows, irrigation and the price of irrigation can determine a farmer's access to land. In these liter-based hydropolitics, the disposition of volumetric water determines the (mal)distribution of territory. In contrast, users of linear water arrive at resolutions to issues of land much more directly, without the intermediate step of irrigation. In fact, such "bounders" do not even measure water per se. Instead, they judge the land contained by aquatic lines. They may sell that land on a per-hectare basis, or, as in the case of Vhimba, they may expend a great deal of effort to secure access to the land. In other words, bounders struggle not over water itself, but over land: the right to inhabit and cultivate the bank of a certain river. Such conflicts can easily be as visceral and violent as those in Vhimba. In well-watered uplands, they are the stuff of revolutions and everyday resistance, a hydropolitics of hectares.

Notes

1. Conducted between 1995 and 1997, this research was assisted by the World Bank, the Center for African Studies (University of California, Berkeley) and a grant from the Joint Committee on African Studies of the Social Science Research Council and the American Council of Learned Societies with funds provided by the Rockefeller Foundation. The Institute on Global Conflict and Cooperation and the MacArthur Foundation supported the writing of this chapter. Joachim Blatter, Helen Ingram, and three anonymous reviewers provided helpful guidance on the manuscript—without, of course, incurring responsibility for its errors or misjudgments.

2. Chiefs bear dynastic titles, such that successors carry the same name as their fathers. The same is true for headmen and kraalheads. As a shorthand, I employ the English colonial terms for these three offices. Readers seeking a discussion of the Ndau terms and their shifts in meaning in the course of the twentieth century are referred to Hughes 1999a.

3. Among secondary sources, Rennie 1973, chap. 3, provides the best account of this period. Among published primary sources, see Don 1997, Leverson 1893, and Mhlanga 1948.

4. Portugal captured the Gaza Nguni king, Gungunyana, in 1895. Britain and Portugal then demarcated the border between their possessions in southeast Africa in 1897, establishing what is now the border between Zimbabwe and Mozambique.

5. In British colonial terminology, this area was a "native reserve." Later, the designation became "tribal trust land" and, after independence in 1980, "communal land." To avoid confusion, I use one term only. Although the natural choice would be "reserve," I employ the American expression for a quite similar phenomenon—"reservation"—to avoid confusion with the botanical reserves established in the 1970s.

6. Hughes (1999a, chap. 3) mentions a number of other processes that contributed to the gradual transformation of headmen's authority from a basis in controlling clients to a basis in controlling land. Behind land alienation, the most important factor was the colonial repression of marriage conventions that subordinated the son-in-law to the father-in-law. The Native Marriages Ordinance of 1908 strictly forbade all forms of "bride service" in favor of a cash bridewealth. By paying cash, sons-in-law freed themselves of most obligations to their fathers-in-law.

7. There is no evidence that newcomers to Vhimba have allocated land to themselves, as Alexander (1993, 120) documents for other parts of Zimbabwe. Nor do elected or appointed government officials allocate land to smallholders here, as distinguished from Dzingirai's (1996) findings in Binga district.

8. According to Alexander (1993, 53–54), land use planning has also failed because steep slopes place much of the Chimanimani mountain area outside the standards of land use planning. In a completely impractical manner, the criteria simply outlaw farming in such areas, many of which are fruitfully cultivated nonetheless.

9. The English words for these offices are particularly misleading when we speak of the period before white conquest. Before land alienation squeezed the African population, rival claimants to the throne could secede from a chiefdom and found their own simply by moving to an "interstitial frontier" (Kopytoff 1987). To prevent this loss of support and subjects, chiefs would have established some form of shared governance among brothers, a pattern still apparent among headmen in Vhimba. Most likely, colonial rule formalized this loose, ambiguous system into the institution of a single titleholder for life (cf. Comaroff 1978).

10. *"Nyika yangu yatorwa nepurazi ... Ndakagara muno maMatwukira ... Ndinotongwa. Itai. Ndoita."*

11. L. S. Jameson, British South Africa Company, to G. B. D. Moodie, March 30, 1892; cited in Olivier 1957, 148.

12. A. H. Duncan, British South Africa Company, to G. B. D. Moodie, October 20, 1892; cited in Olivier 1957, 150.

13. An anomalous farm, Glencoe, actually split the Ngorima Reserve, into two parts, Number 1 and Number 2. The lower section of that farm was reclassed as African land in 1925 (Palmer 1977, 263).

14. *"Takanga tine nzimbo hombe. Takaitorerwa. Tasara nekaplace kadiki-diki."*

15. Riddell 1979, 20–21. The relevant land categories are the Natural Regions II and III, government designations based on rainfall. See also Moyana 1984, Moyo 1995, and Moyo et al. 1991.

16. *"Munogona kugara pamunoda";* recalled in 1996 by a headman's nephew.

17. Numerous authors have addressed the flaws in Zimbabwe's resettlement program of the 1980s (Jacobs 1983, 1991; Kinsey 1983; Potts and Mutambirwa 1997). Where the program worked according to plan, strict criteria barred from resettlement farmers who engaged in dry-season migrant labor and those whose farm productivity was not exemplary. Where spontaneous action overwhelmed resettlement officers, smallholders already living adjacent to the resettlement area and their relatives and clients grabbed the land.

18. As shown in aerial photographs available from the Office of the Surveyor General, Harare.

19. The farmer, who made his appeal at a gathering in early 1997, has not pursued the issue. The compromise on the boundary of Hayfield B appears to have held.

20. *"Yaigarwa kare. Mafuta akafirei-wo. Maparara akafirei-wo."*

21. *"Ngatiedze kuita zvido zvemuridzi wenzvimbo ino, David Muhanyi. Kugara nevamwe zvakanaka munharaunda. Chengetai zvipfuyo zvenyu."*

22. Headmen tended to discriminate against Mozambicans, as opposed to Zimbabwean internal migrants, in the allocation of land such that the former received plots in insecure, disputed areas. Mozambicans, for their part, tended to bargain less and demand choice parcels to much a lesser extent than their Zimbabwean counterparts. Hughes (1999a, chap. 4) discusses the reasons for this discrepancy.

23. Chirwa and Mavi (below in text) are pseudonyms, used to protect the anonymity of the actual settlers.

24. *"Vanoda kurima* boundary." *"... kuti vaite* line *rimwe chete."* (People in Vhimba often used these English words when describing disputes over demarcation.)

25. M. himself was spending most of his time as migrant laborer in South Africa.

26. "Report to the Parks and Wild Life Board concerning the Protection of the Haroni-Rusitu (Lusitu) Botanical Reserves," February 1995, 1.

27. *Tinokumbirawo, Cde. Minister ... nzvimbo dzedu dzakakosha* (tribal monuments) *dzakadai nge.... Gwasha rekwaNyakwaa kwaVhimba ... Tingachengete idzi nzvimbo, nyika yedu inozogara yakanaka nekuramba yakapuwa donhodzo ndiMusiki.*" Speech by the spokesperson for the Ngorima lineage at the installation of Chief Peter Ngorima, at Rusitu Mission, Chimanimani District, Zimbabwe, March 31, 1995.

28. The committee's early activities fell within the scope of Zimbabwe's Communal Areas Management Programme for Indigenous Resources, an internationally acclaimed effort toward community-based management of natural resources. For more on Vhimba's involvement and de facto secession from that program, see Hughes 1999a, chap. 5.

29. I was able personally to circulate a version of the map to which I had added the dates of land allocation for each of the families.

References

Alexander, Jocelyn. 1993. "The State, Agrarian Policy and Rural Politics in Zimbabwe: Case Studies of Insiza and Chimanimani Districts, 1940–1990." Ph.D. diss., Oxford University.

Beach, D. N. 1986. *War and Politics in Zimbabwe, 1840–1900*. Gweru, Zimbabwe: Mambo Press.

Burrows, Edmund H. 1954. *The Moodies of Melsetter*. Cape Town and Amsterdam: Balkema.

Comaroff, John L. 1978. "Rulers and Rules: Political Processes in a Tswana Chiefdom." *Man* 13 (new series): 1–12.

Don, J. B. 1997. *Journal* (1891–1892), reprinted (D. N. Beach, ed.) in *Heritage of Zimbabwe* 16: 6–31.

Dzingirai, V. 1996. " 'Every Man Must Resettle Where He Wants': The Politics of Settlement in the Context of Community Wildlife Management Programme in Binga, Zimbabwe." *Zambezia* 23, no. 1: 19–30.

Godwin, Peter. 1996. *Mukiwa: A White Boy in Africa*. New York: Harper Collins, 119–120.

Hughes, David M. 1996. "When Parks Encroach Upon People: Expanding National Parks in the Rusitu Valley, Zimbabwe." *Cultural Survival Quarterly* 20, no. 1: 36–40.

Hughes, David M. 1999a. "Frontier Dynamics: Struggles for Land and Clients on the Zimbabwe-Mozambique Border, 1890–1997." Ph.D. diss., University of California, Berkeley.

Hughes, David M. 1999b. "Refugees and Squatters: Immigration and the Politics of Territory on the Zimbabwe-Mozambique Border." *Journal of Southern African Studies* 25, no. 4: 533–552.

Jacobs, Susie. 1983. "Women and Land Resettlement in Zimbabwe." *Review of African Political Economy* 27/28: 33–50.

Jacobs, Susie. 1991. "Land Resettlement and Gender: Some Findings." *Journal of Modern African Studies* 29, no. 3: 521–528.

Kinsey, B. H. 1983. "Emerging Policy Issues in Zimbabwe's Land Resettlement Programmes." *Development Policy Review* 1: 163–196.

Kopytoff, Igor. 1987. "The Internal African Frontier: The Making of an African Political Culture." In Igor Kopytoff (ed.), *The African Frontier: The Reproduction of Traditional African Societies*, pp. 3–81. Bloomington: Indiana University Press.

Latham, C. J. K. 1970. "Dzimbadzemabgwe." *Native Affairs Department Annual* (Rhodesia) 10, no. 2: 24–30.

Leverson, J. J. 1893. "Geographical Results of the Anglo-Portuguese Boundary Delimitation Commission." *Geographical Journal* 2: 505–518.

Mhlanga, Wilson. 1948. "The Story of Ngwaqazi, 2: The History of the Amatshangana." *Native Affairs Department Annual* (Southern Rhodesia) 25: 70–73.

Minutes of meeting of Vhimba Area Development Committee, Vhimba, November 3, 1996.

Moore, Donald Shearer. 1995. "Contesting Terrain in Zimbabwe's Eastern Highlands: The Cultural Politics of Place, Identity, and Resource Struggles." Ph.D. diss., Stanford University.

Moyana, Henry V. 1984. *The Political Economy of Land in Zimbabwe*. Gweru, Zimbabwe: Mambo Press.

Moyo, Sam. 1995. *The Land Question in Zimbabwe*. Harare, Zimbabwe: Southern Africa Political Economy Series.

Moyo, Sam, Peter Robinson, Yemi Katerere, Stuart Stevenson, and Davison Gumbo. 1991. *Zimbabwe's Environmental Dilemma: Balancing Resource Inequities*. Harare: Zimbabwe Environmental Research Organisation.

Olivier, S. P. 1957. *Many Treks Made Rhodesia*. Cape Town: Howard B. Timmins.

Palmer, Robin. 1977. *Land and Racial Domination in Rhodesia*. Berkeley and Los Angeles: University of California Press.

Potts, Deborah, and Chris Mutambirwa. 1997. "'The Government Must Not Dictate': Rural-Urban Migrants' Perceptions of Zimbabwe's Land Resettlement Programme." *Review of African Political Economy* 74: 549–566.

Rennie, John K. 1973. "Christianity, Colonialism and the Origins of Nationalism among the Ndau of Southern Rhodesia." Ph.D. diss., Northwestern University.

"Report to the Parks and Wild Life Board concerning the Protection of the Haroni-Rusitu (Lusitu) Botanical Reserves," February 1995: 1.

Rhodesia Act 5 of 1979, Land Tenure (Repeal), p. 116.

Riddell, Roger. 1979. "Prospects for Land Reform in Zimbabwe." *Rural Africana* 4/5: 17–31.

Sinclair, Shirley. 1971. *The Story of Melsetter*. Salisbury: M.O. Collins.

Sleigh, R. W. 1976. "Survey and the Surveyors." *Rhodesia Science News* 10, no. 1: 3–4.

Speech by the spokesperson for the Ngorima lineage at the Installation of Chief Peter Ngorima, Rusitu Mission, Chimanimani District, Zimbabwe, March 31, 1995.

Thongchai, Winichakul. 1994. *Siam Mapped: A History of the Geo-Body of a Nation*. Honolulu: University of Hawaii Press.

III
Lessons for Theory, Research, and Governance

11

Perspectives from the Districts of Water and Power: A Report on Flows

Richard Perry

Water Politics in a World of Flows

In this chapter, we return to the central conceptual issues raised in the two introductory chapters: the challenges posed by water to classical models of territorial governance under the nation-state system. Although water has played an important, if insufficiently noted, historic role in the construction of nation-state boundaries and infrastructures, the fluidity of water challenges the stable, steady-state logic of territorial governance of the nation-state.

Although we who have been collaborating on this volume have thought of it all along as a book about water, and even though all of the case studies collected here touch in one way or another upon water, the more profound claim underlying our work is that the dynamics of water may show us ways to approach the most important issues of governance of our time. Water, we argue, is good to think with.

As I discuss further in the next section, the politics of water brings into sharp focus what is too often forgotten in environmental scholarship: that the territorial nation-state framework of modern governance is an artifact of quite recent invention. The territorial nation-state form, as it was constructed in the context of ideological conflict and decades of warfare in early modern Europe, accomplished the remarkable feat of at the same time nationalizing nature and naturalizing the nation-state.

Thus, the received logic of national/natural resource management—and the metaphysics of the nation-state upon which this logic depends—is instructively confounded by diverse problems of transboundary watersheds, river basins, bio-regions, offshore fisheries, and environmental degradation of water basins such as lakes and seas, like those examined in this volume. We see that water has been central to the historical organization of the modern nation-state form, both as territorial boundary of the nation-state and as its primary infrastructure for transport and communication. I argue that water is now no less revealing of the nation-state's erosion, its transformation under the conditions of late or post-modernity, or both. The fluidity of water politics holds the boundaries of nation-states in solution (perhaps in dissolution these days) and, taken together, the studies here point toward new forms of sub-, supra-, trans-or paranational governance whose outlines may just be discernible as they precipitate out of this solution.

"Globalization," as noted in chapter 1, is the buzzword of our time, a one-size-fits-all noun for a wide array of phenomena that distinguish our moment at the *fin de millennium*. Yet "globalization" is a noun whose unreflective commonsensical ubiquity seems more often to stanch the flow of inquiry than to facilitate it. Nevertheless, if "globalization" has any core meaning across all the contexts to which it is applied, globalization means "flows" (see Castells 1996)—flows of capital, goods, services, resources, images, people—and of course, as is noted in chapter 1, water serves as the root metaphor for all the ways in which these flows are conceptualized: capital flows, streams of commerce, market stagnation, offshore labor pools, waves of migrants, floods of information, hegemonic cultural inundation, and so on. It is the future history of the practices and politics and networks of such flows upon which the hopes for the development of a global civil society rest (Lash and Urry 1994; Appadurai 1996; Hannerz 1996; Basch, Schiller, and Szanton Blanc 1994).

Another way of grasping the commonality of all of the above-cited flows of globalization is to see these flows that are reflected in watery images as directional forces, as varied and variable forms of power. Yet the key point here is that this is not simply the classical coercive state regulatory power of modern political governance, but rather the "capillary" power of myriad practices whose specific rationalities and effects Michel Foucault termed "governmentality" (see chapter 6 and Burchell, Gordon, and Miller 1991). In his last years, Foucault repeatedly turned to the spatial and geographic dimension of power and its effects. As he said,

a whole history remains to be written of *spaces*—which would at the same time be the history of *powers* (both these terms in the plural)—from the great strategies of geo-politics to the little tactics of the habitat.... Anchorage in a space is an economico-political form which needs to be studied in detail. (1980, 149)

Foucault is the great theorist of the specific, the particular, the local; for him and for the many other recent scholars who have carried his analysis further (see chapter 6; Scott 1998; Ong 1999; Ferguson 1998), power is never the absolute, totalizing, top-down ordering of classical theory.[1] Rather, any specific exercise of power summons into being its own local counterformations. Power, one might say, is conceptualized hydraulically, as constant ebbs and flows of agency and resistance.

To situate Foucault's notion of power/knowledge alongside such a "hydraulic" conceptualization of the effects of specific "governmental rationalities" enables us to approach contemporary transformations in the nation-state with fresh eyes. Anthony Giddens has defined the nation-state as "a bordered power-container ... [as] the pre-eminent power-container of the modern era" (Giddens 1985). There is a double sense of "contain" in Giddens's phrase, for we see that the discourses and practices of modern state governance are conceived as power effects whose fields of flow are both localized within the territorial vessel of a nation-state and "contained," constrained, and delimited by its borders. Even in the case of overt warfare or other interstate conflict, we see not simply a clash of raw forces but rather a clash of "power-containers" asserting themselves as such. The confusions in the Abkhaz-Georgian conflict, as shown in chapter 8, illustrate this principle: Just as each of the two parties is a new or even a quasi state seeking through force of arms to emerge and to be recognized as a legitimate "power-container" with control over its "own" "natural" territory.

Nation-states, therefore, are, and have always been, historically emergent constructs of their own ideological and practical efficacity. Yet it is of the utmost importance to note that to acknowledge this fact is not at all the same thing as naively denying the actuality of states. Rather, it is to recognize the contingent and artifactual nature of state formations and to broaden the channels of investigation, to open the sluice gates of inquiry into the specific local effects of regulatory and managerial power and of their oppositional formations, and perhaps to enable ourselves to envision alternatives to the present conceptual status quo.

My goal in the next section is a sort of "archeological" (in the sense of Foucault 1972) excavation of the sedimented meanings of water, power, and governance against the background of the rise, the flourishing, and the contemporary transformation of the nation-state form.[2] I argue that the received models for governing water and its associated geographic formations, especially in border regions as we have seen in the studies presented in this collection (watersheds, river valleys, bioregions, etc.), exhibit an incoherence that flows from the implicit assumption of a natural homeostatic ontology of nation-states, from what Liisa Malkki (1992) has called the "sedentarist metaphysics" of the modern nation-state form. Classic political theory has long remained blindered to this contingent constructedness of nation-states, rendering itself incapable of imagining alternate orders of either politics or nature. In the subsequent section, I consider the more recent transformations of the modern nation-state form in the new world of "flows," especially in regard to environmental governance, associated with globalization/glocalization in the period called "postmodern."

From the Wellsprings of Civilization to the National Order of Things

By the Waters of Babylon: Civilization and Wet Regions

Western civilization, according to the autobiography that it has written for itself, was born between two watercourses. According to ancestral sacred texts, our Genesis was in Mesopotamia (Greek for a region between rivers), our primal Eden was somewhere in the Fertile Crescent. Other civilizational origin tales are similarly located in riparian zones; the Nile, the Indus, the Hwang Ho and Yangtse are all the names not only of rivers but also of originary civilizations that were to spread outward from their riverine beginnings to found the major cultural regions of the world.

In each of these riparian regions, the presence of available water for increasingly efficient food production through sedentary agriculture is thought to have enabled greater population density, the formation of urban centers, socioeconomic specialization, and ultimately the development of literacy and the accretion of collective forms of knowledge that we know as civilization. This is a standard story of civilizational thinking (for a version of this narrative embedded in a specifically environmental history of human civilization, see Ponting 1991, chap. 4). It is less often noted, however, that such civilizational narratives are all *regional* stories in which the development of specific civilizations is rooted in specific riverine regional environments. Even the runners up in the great historical competition among civilizations were closely linked to water: the Mississippi/Ohio River Valley mound builders, the Mayans of the lowland rainforests, and the trading cultures of the Congo and the Niger, among many others.

The first sentence of a brochure for a 1999 conference on "Civilizational Thinking" at the University of California, Santa Cruz, asks: "How are regions made and unmade?" It goes on to define "civilizational thinking" as "the idea that knowledge and societies are organized into regional legacies." Colin Ward's useful overview of water politics (1997) includes a chapter entitled "Hydraulic Societies and Regional Hopes," in which he connects grand historical theories of civilizational development to the nineteenth- and twentieth-century foundations of regional planning (and to the Tennessee Valley Authority and the California Aqueduct), through the intellectual lineage that runs from the French physiocrats, through the Russian anarchist geographer Peter Kropotkin, to the Scottish theorist of bio-regions, Patrick Geddes, and to Geddes' American follower, Lewis Mumford, and the foundation of regional planning theory in North America.

One common assessment measures any civilization's historical progress according to the degree of its mastery of the watery conditions of its birth, that is, by its increasing control of water, as evidenced by the engineering of canals and dams and other regional waterworks. Many earlier or even "lost" civilizations are known to contemporary scholarship largely by virtue of archeological traces of their irrigation practices. Indeed, it is telling that early modern astronomers proposed the existence of kindred life forms on the planet Mars based upon what they perceived through their early telescopes to be irrigation canals.

Later generations of Europeans saw the presence and, in some cases, continued functioning of Roman aqueducts across Western Europe as abiding, still-visible relics, reminders of an earlier advanced civilization that had survived the so-called Dark Ages between the collapse of the Roman Empire and the early modern era. It is often noted that many regions in Europe enjoyed better access to water for irrigation, drinking, and plumbing under the Romans than they did until the eighteenth, nineteenth, or even the twentieth centuries—that is, when this region had regained its earlier level of civilizational advancement.

In the biblical book of Genesis, the supervention of the order of the Logos over primordial Chaos at the inception of human earthly time is marked by the imposition of land upon the waters and of Adam and Eve's primal Edenic garden upon the land. One classic essay in the history of environmentalist thought traces this history of the subjugation of Nature in the ideology of the monotheistic West; it argues that

Christianity, in absolute contrast to ancient paganism and Asia's religions (except, perhaps, Zoroastrianism), not only established a dualism of man and nature but also insisted that it is God's will that man exploit nature for his proper ends ... [and] we shall continue to have a worsening ecologic crisis until we reject the Christian axiom that nature has no reason for existence save to serve man. (White 1966, 26, 29)

The Peace of Westphalia that concluded the internecine Christian religious strife of the Thirty Years War in 1648 heralded a new theology, ultimately a new scientific cosmology of nature and of reason. The midseventeenth century saw the beginning of a new world order grounded in a shift in the relation between nature and the human society. Rather than the Earth and its features being at the far end of the "great chain of being" from the Logos, as in the Ptolemaic cosmology of the medieval period (see Lovejoy 1948), the early modern era of seventeenth- and eighteenth-century Europe and North America witnessed the epistemological revolution of Bacon, Hobbes, Locke, Newton, Rousseau, Montesquieu, Jefferson, Kant, and Hegel.

This early modern period from the late seventeenth to the early nineteenth centuries marked a shift from the medieval grounding of the social order in norms derived from scripture and kinship to a new territorial vision of the modern nation-state. The source of order moved from otherworldly divinity to an identification of the Logos with nature's reason, with the Newtonian laws of the material world of nature. Early scientists, like Newton, saw themselves as Christians doing God's work as they sought to derive natural law not only from Scripture, but also by their close reading of the Book of the World. Benedict Anderson, the leading contemporary student of the nation-state, cites the influential arguments of Rousseau and Herder that "climate and 'ecology' had a constitutive impact on [national] culture and character" (1991, 60).

It was Montesquieu who, in his 1751 *De l'esprit des lois*, proposed a theory of social-ecological regions, natural/national reason, and the government of laws. Montesquieu argued that local geography and climate combined to produce different national characters and that there must therefore be laws appropriate to each nation's "nature." Temperate climes and island or mountainous geographies, like those of England and Switzerland, are conducive to democratic government; hot countries like India or flat lands like Poland are prone to despotism.

Montesquieu was hardly an isolated eccentric; indeed, he was the most influential eighteenth-century proponent of constitutional government and the primary theorist of the tripartite separation-of-powers model that first achieved realization in the U.S. Constitution. Thomas Jefferson declared himself Montesquieu's disciple and was the thinker who most effectively carried Montesquieu's ideas into practice, and not only in the U.S. Constitutional Convention of 1787.

Jefferson's 1800 work, *Notes on Virginia*, was a striking example of Montesquieu's political ecology. Jefferson observed that the different climates and land forms of the thirteen states produced distinct local characters, from hot-blooded Southerners to laconic New Englanders, and that each of the new states ought therefore to have a set of laws appropriate to its character. Jefferson's exposition in *Notes on Virginia* proceeded in the manner of a scientific proof. He began by describing the geographic formations of Virginia: the low-lying coastal wetlands of the Tidewater region, the elevations and watercourses of the Piedmont and the western mountains. He then described the flora and fauna of the regions, and then the three human races—the native tribes, the slaves of African descent, and the white citizens—and their patterns of settlement, forms of habitation, and relations to one another. Finally, Jefferson proceeded to "derive" the constitution and legal code of Virginia (much of which he had himself drafted according to this organicist regional logic) from the specific "natural" facts of Virginia's land, climate, flora, fauna, and peoples.

On the European continent, Jefferson's younger contemporary, German philosopher G. W. F. Hegel, proposed at almost the same time a less empirical theory of the modern nation-state, although the most immediate French influence on Hegel was not Montesquieu but Napoleon Bonaparte. Napoleon's advance across Europe brought along with it the metric system, a new calendar whose months were named for the season's weather, and a new model of the rule of law in the Code Napoleon. For Hegel, the conquest of his own region of Germany, along with most of the rest of Europe, by Bonaparte—whom he called "the Messenger of the World-Spirit"—signaled a culmination of human history in the convergence of the rational and the real, of the ideal and the material, of Nature and Reason in the form of the modern nation-state.

Natural Geographic, the National Order of Things

The rise of the nation-state form in early modern Europe has been discussed here at length precisely because the nation-state must be regarded as one of the most successful inventions in modern human history. Indeed, the point is that the nation-state form has been so successful that the very fact that it is an invention, and one of relatively recent provenance, is so generally forgotten.[3] It is so taken for granted that it goes about its business tacitly bounding and structuring the lives and experiences of the greater part of humanity; it is taken as natural, as simply the national order of things.

Thus the territorial nation-state's patchwork patterned map of the surface of the Earth has, since the seventeenth century, come to be seen simply as part of the natural/national order of things according to which the seven seas, the five continents, the four Linnean races of humankind (identified with the four quadrants of the Cartesian grid coordinates of the compass: the North-European-white, the South-African-black, the East-Asian-yellow, the West-American-red) all have their natural, geographically ordained place on the racial-spatial map of the Earth. The classical theorists of the nation-state, such as Montesquieu, Jefferson, and Hegel, all grounded their visions of the modern *Rechtstaat* in this natural order and its "natural laws." These theorists' models have explicitly or implicitly undergirded the post-Westphalian international order of territorially sovereign nation-states and its political theory.

If we are to grasp the full significance of the nation-state form for environmental governance, we must understand what Anderson calls the "grammar" (1991, xiv) of the nation-state and its characteristic "modular" elements (1991, 4). I argue that the grammar of the nation-state is constituted by its "developmental" narrative of origins, rise, and progress, typically exemplified in the nation-state's progressive mastery of "its" territory, of "its" portion of "nature," of its own(ed) "natural resources."

In a widely cited essay on nationalist territorialization, Liisa Malkki offers a "schematic exploration of taken-for-granted ways of thinking about identity and territory that are reflected in ordinary language, in nationalist discourses, and in scholarly studies" (1992, 25). In a section of her essay entitled "Maps and Soils," Malkki questions the "commonsense ideas of soils, roots, and territory built into everyday language and often also into scholarly work, ... [ideas whose] very obviousness makes them elusive as objects of study" (1992, 26). Malkki cites a passage from Ernest Gellner's 1983 book *Nations and Nationalism*, in which Gellner describes two hypothetical maps, one drawn up before the rise of the nation-state form and the other after. As Gellner says, the "first map resembles a painting by Kokoschka ... a riot of diverse points of color ... such that no clear pattern can be discerned in any detail." Gellner's second map, an

ethnographic and political map of an area of the modern world ... resembles not Kokoschka, but, say, Modigliani. There is very little shading; neat flat surfaces are clearly separated from each other, it is generally plain where one begins and another ends, and there is little if any ambiguity or overlap. (Gellner 1983, 139–140, quoted in Malkki 1992, 26; see also the discussion of Gellner in Perry 1995)

Malkki observes that Gellner's second map is

much like any school atlas with yellow, green, pink, orange and blue countries composing a truly global map with no vague or "fuzzy spaces" and no bleeding boundaries. The national order of things ... also passes as the normal or natural order of things. For it is self-evident that "real" nations are fixed in space and "recognizable" on a map. One country cannot at the same time be another country. The world of nations is thus conceived as a discrete spatial partitioning of territory; it is territorialized in the segmentary fashion of the multicolored school atlas. (Malkki 1992, 26)

Malkki illustrates the "naturalizing" effects of the characteristic deployment of "specifically botanical metaphors"—soil, roots, ethnic stocks, and branches—in the service of what she calls the "sedentarist metaphysics" of territorial nationalism. There is an evident etymological connection between "natal" concepts of a common organic origin and the concepts of "nation," "nature," and the "native."

Malkki shows that this metaphysics has particular implications for those who are seen as "natives," who are thought to be most firmly "rooted in place" in the territory of a specific region, that is, who are "autochthonous" or "indigenous." Others, she points out, frequently take "natives" as symbols of the nation-state. Recall here, for example, the appearance of "Indian head" coins in the late-nineteenth-century United States just as the greatest of the genocidal campaigns against the Plains tribes were being conducted, or the 1960s environmentalist television campaign depicting the "crying Indian" who was weeping, evidently, not because of the displacement and impoverishment of the first peoples of North America but because he was seen to be especially sensitive to the environmental degradation of the entire continent—the crying Indian was crying on behalf of the rest of us. As Malkki observes, it is a consequence of the nation-state's ideological need for rootedness that the "[t]he 'natives' are ... [spatially] incarcerated in primordial bioregions and thereby retrospectively recolonized" in this sort of environmentalist discourse (Malkki 1992, 30).[4]

Historically, through the nineteenth and twentieth centuries, a key modular form of the nation-state's grammar of development has been the control and damming of its hydrological "assets." The narrative of nation-state development and modernization has increasingly been symbolized by great waterworks. Much as classic civilizational thought was linked to rivers—from the Nile to China's Three Gorges—twentieth-century nation-state modernization has been closely identified with grandiose dam projects. The nation-state's identification of its identity and its progress with water is exemplified in grandiose water/power management schemes such as the Tennessee Valley Project, Sudan's

Gezira Scheme, the Aswan High Dam, and the contemporary immense Three Gorges undertaking in the People's Republic of China. Interestingly, in many cases the fact that the construction of the waterworks would disrupt large regions of the "natural" environment and displace entire populations of "natives" only seemed to strengthen the project's appeal as testimony to the power and unity of the nation-state (the Grand Coulee Dam is one North American example, and the Quebec nationalist party government's enormous hydroelectric dam project on Cree ancestral lands is another) (Ignatieff 1993; see also chapter 3).

Benedict Anderson has argued that, although the ideology of the nation-state may have been conceived in Old Europe, the modular structures that have characterized the nation-state form were developed in the overseas context of colonial administration and socioeconomic development. Some of the greatest nineteenth- and early-twentieth-century hydrological enterprises—the Suez and Panama Canals, for example—were very much part of the subjugation of colonized regions for the benefit of colonizers. A fascinating study by Victoria Bernal examines the "Gezira Scheme [in Sudan],... the largest centrally-managed irrigation project in the world" (1997, 447). It was built by the British for cotton production and began operation in 1925. As Bernal notes,

while the Gezira Scheme was not very successful in terms of economic performance, it succeeded in the colonial era as a monument to economic modernization and the values of rationality, discipline, and order and has continued to function as a potent symbol of progress and state power until today. (1997, 448)

Colin Ward has traced the importance of the enormous dam-building projects undertaken by the U.S. government in the 1930s: Hoover Dam, Grand Coulee Dam, and "the supreme example of the use of water engineering" in the Tennessee Valley Authority (Ward 1997, 38). The Soviet Union undertook no less massive dam projects, and both the United States and the Soviet Union competed to provide expertise and financing for the largest dam of the last generation, Egypt's Aswan High Dam. Today, China is undertaking an even more massive dam project, the Three Gorges Dam, which is projected to displace at least 1.1 million people (Ward 1997, 36).

The great dam project, which symbolizes the power and unity of the nation-state, provides an apt symbol for Giddens's definition of the nation-state as a "bordered power-container." According to calculations by the National Aeronautics and Space Administration (NASA), the redistribution of large masses of water by dam projects has shifted the angle of the Earth's axis by approximately two feet since 1950 (cited in "Harper's Index" 1997, 13, 87).

The Fluid Meanings of Nature/Water in a World of Flows

The narrative of nation-state development that I have been discussing is the conceptual framework that Paula Garb and John Whiteley refer to in chapter 8 when they note that the national security concerns that structure the case of the Inguri Dam complex "are issues of the modern world, not the postmodern." This is surely a correct reading. Garb and Whiteley contrast these "modern" issues with other studies' concerns with a specific ecosystem, which, they argue, is "a postmodern conceptualization" (chapter 8).

Yet one implication of my historical account of the nation-state form is that a focus on the eco-region might be read as either postmodern or premodern, or perhaps both. One interesting implication of Garb and Whiteley's reading, found in the other case studies in this volume as well, is that the actualities of contemporary environmental politics exhibit features that the standard modernization story (summarized in chapter 2) would classify as premodern, modern, and postmodern. This fact is not, perhaps, quite as perplexing as it seems at first. For if, as I have argued above, the modern nation-state-based world order has always been as much ideological as actual, then it is not entirely surprising that the actuality has not always conformed to the ideology, however difficult this may be for political theorists to recognize. The persistence of "premodern" traits, the emergence of "postmodern" features: both of these facts are consistent with the assertion by the sociologist of science, Bruno Latour, that "we have never been modern" (the title of Latour 1993).

Put in different words, the very fact that we expect premodernity, modernity, and postmodernity to manifest themselves as discrete stages that follow and supplant one another is itself indicative of a quintessentially modernist paradigm of thought: that history should follow the stages that modernization theory has foreordained for it. Why should premodern "customary" kinship-based societies *necessarily* be superseded by modern "rationalized" territorial nation-states that in turn *nec-*

essarily shrivel up and blow away under the advancing flows of "hyper-rational" globalization? Why should we be surprised to find what we have found in the studies in this volume: that virtually all of these forms of governance can be identified, not in the theoretical imagination, but in actual practice in diverse local and regional contexts today?

Development scholar James Ferguson (see Ferguson 1990) has recently pointed out that such classical evolutionary stage narratives, despite the fact that the nineteenth-century social thinkers who conceived them were drawing upon Darwinian models (see here Marx's dedication of *Capital* to Charles Darwin), turn out not to work well even for biological history (as Stephen Jay Gould has been pointing out for years: Dinosaurs and saber-toothed tigers are indeed no longer with us, but the most "primitive" early life forms such as bacteria might well be adjudged the most enduring and successful right down to the present moment, and we relatively recently arrived homo sapiens ought not be overhasty in proclaiming ourselves the last and highest stage of life—the protozoans may very well write our eulogies). So why, Ferguson asks, should we expect that such evolutionary stage models will be adequate to describe developments in human sociopolitical history, such as globalization?

Anthony Giddens has defined globalization as "the intensification of worldwide social relations which link distinct localities in such a way that local happenings are shaped by events occurring many miles away and vice versa" (1990, 64). Benjamin Barber's 1995 book *Jihad v. McWorld* is the best-known formulation of what many analysts see as the two alternative and opposing paths available to us in our "postmodern," globalizing moment. On the one hand are the allegedly global homogenizing effects of planetary commerce, especially as conducted by transnational corporations. In this view, the leveling effects of global marketing and consumption practices are reconstructing all of humankind into a universe of effectively fungible "McCitizens" and "McConsumers" who work at "McJobs" and ingest increasingly indistinguishable "McFood" as we sway to the rhythms of global MTV.

It should be stressed that Barber's ironically critical view of "McWorld" is not an isolated position but is nevertheless something of a minority viewpoint among Western academic theorists. Indeed, the majority have tended to see the homogenizing tendencies of global flows of products, services, capital, and cultural imagery as the final act in modernity's emancipation of the individual from the constraints of the local. This utopian vision of globalization is rather like the ultimate convergence of the rational and the real at the end of history (Fukuyama 1991), that is, as the final arrival of Hegel's world spirit in the form of global civil society that has outgrown all but the minimal state control of a "nightwatchman government."[5]

The alternate face of Barber's dichotomous view of contemporary globalization is represented by what he calls "Jihad." In this usage, "Jihad" stands for a "holy war" waged in the name of any premodern, or even antimodern, local, or regional particularism, "fundamentalism," or essentialism; it stands especially for any war waged against the Enlightenment universalism embodied in the modern secular state, against its government of laws, and against the homogenizing, deracinating effects of "McWorld's" global capitalism.[6] Historian Tony Judt describes this as "a bizarre resurrection of the ghosts of particularism" (1994, 44). Michael Ignatieff argues in his 1993 book *Blood and Belonging: Journeys in the New Nationalism* that "the key narrative of the new world order is the disintegration of nation-states into civil war; the key architects of that order are warlords; and the key language of our time is ethnic nationalism" (1993, 5).

It is instructive that both formulations, Jihad as well as McWorld, foresee a transformative fragmentation of modern nation-state sovereignty. Both sides are susceptible to an overwrought Chicken Little reaction: "The state is falling! The state is falling!" Yet no matter whether this exclamation is fearful or hopeful, it is surely overstated. The nation-state form is undergoing a diverse ensemble of changes, but rumors of its demise are much exaggerated.

The nation-state-based cartography of modernity is being reconfigured, not erased. What Malkki described as the "truly global map with no vague or 'fuzzy spaces' and no bleeding boundaries, ... the world of nations ... conceived as a discrete spatial partitioning of territory ... territorialized in the segmentary fashion of the multicolored school atlas," is being transformed, although it is the nation-state that remains the key entity whose transformation is reshaping the globe (Malkki 1992, 26). The school atlas map's stable and neatly bounded fields of color are leaching into one another across their border regions, due in large measure to increasing transnational flows of capital, resources, people, images, and even regulatory power and practices of governance; these are the developments that have been producing what has been variously called "perforated," "graduated," or "dispersed" sovereignty (chapter 2; Ong 1999; Santos 1995). At the same time, formerly existing "power-containers" such as the Soviet Union have been fractionating into ethno-regions, a process in which each fragment claims for itself, like a holo-gram, its own holistic nation-state identity and sovereignty, seeking to be recognized by other power-containers as a power-container in its own right.

These apparently countervailing tendencies toward increasing socio-economic globalization on the one hand and intensifying localism on the other are captured as two facets of a single dynamic dialectical process in the concept of "glocalization" (see chapter 1 and Robertson 1998). As Giddens observes,

> this is a dialectical process.... *Local transformation* is as much a part of global-ization as [is] the lateral extension of social connections across time and space.... The outcome [of factors such as world money and commodity markets] is not necessarily, or even usually a generalized set of changes acting in a uniform direc-tion, but consists in mutually opposed tendencies. The increasing prosperity of an urban area in Singapore might be causally related, via a complicated network of global economic ties, to the impoverishment of a neighborhood in Pittsburgh whose local products are uncompetitive in world markets....
>
> The development of globalized social relations probably serves to diminish some aspects of nationalist feeling linked to nation-states (or some states) but may be causally involved with the intensifying of more localized nationalist sen-timents. In circumstances of accelerating globalization, the nation-state has become "too small for the big problems of life, and too big for the small prob-lems of life." At the same time as social relations become laterally stretched and as part of the same process, we see the strengthening of pressures for local auton-omy and regional cultural identity. (Giddens 1990, 64–65, italics in original)

Far more useful, therefore, than the Chicken Little exultations/lamen-tations that "the state is falling!" that one encounters in so much of the literature on globalization are the recent works that undertake a close examination of both the ideologies and concrete practices of state gov-ernance and of their transformation under what many writers have called the "postmodern" turn.[7] James C. Scott's book *Seeing like a State* is an exemplary overview of the modern state as an ensemble of specific practices of governmentality, practices that seek to render "legible" to the state's gaze the features of its territorial geography and its population in a quite specific fashion. As Scott says in the first sentence of the first paragraph of chapter 1 (entitled "Nature and Space"), "[c]ertain forms of knowledge and control require a narrowing of vision" (Scott 1998, 11). It is just this narrowing or "simplification" that

in turn, makes the phenomenon at the center of the field of vision more legible and hence more susceptible to careful measurement and calculation. Combined with similar observations, an overall, aggregate, synoptic view of a selective reality is achieved, making possible a high degree of schematic knowledge, control, and manipulation. (1998, 11)

Such simplifying, reductionist practices of "schematic knowledge, control, and manipulation" clearly manifest what Scott calls the state's "abstracting utilitarian logic" (1998, 13). They nicely exemplify Foucault's notion of the play of "power/knowledge" in practices of "governmentality." I suggest that this abstracting logic of state governance also represents what Giddens has called the "disembedding" that characterizes late modernity. Further, I argue it is precisely this abstracting, disembedding tendency in modern practices of nation-state governance that has enabled, perhaps compelled, the modern order to transform itself from within, according to the dictates of its own disruptive logic.

Giddens defines "disembedding" as

the "lifting-out" of social relations from local contexts of interactions and their restructuring across indefinite spans of time-space.... The image evoked by disembedding is better able [than are classical evolutionary models of societal development] to capture the shifting alignments of time and space which are of elementary importance for social change in general and for the nature of modernity in particular. (1990, 21–22)

Giddens distinguishes two types of disembedding mechanisms: "the creation of *symbolic tokens*" and "the establishment of *expert systems*" (1990, 22, italics in original). The former mechanism comes into play when entities are, in Scott's words quoted above, "simplified" and rendered "schematic" for purposes of knowledge, control, and manipulation. This is the sort of process referred to in chapter 2 as the "monetarizing" of water. Scott gives the example of the "vocabulary used to organize nature" (1998, 13). Here we see, says Scott, that

in fact, utilitarian discourse replaces the term "nature" with the term "natural resources," focusing on those aspects of nature that can be appropriated.... [This] logic extracts from a more generalized natural world those flora or fauna that are of utilitarian value (usually marketable commodities) and, in turn reclassifies those species that compete with, prey on, or otherwise diminish the yields of the valued species. Thus plants that are valued become "crops," the species that compete with them are stigmatized as "weeds," and the insects that ingest them are stigmatized as "pests." Thus, trees that are valued become "timber," while species that compete with them become "trash" trees or "underbrush." The same logic applies to fauna. Highly valued animals become "game" or "livestock," while those animals that compete with or prey upon them become "predators" or "varmints." (1998, 13)

This logic is not, of course, simply a matter of classificatory pedantry; it is also profoundly productive of environmental transformation. As Iain Boal notes, "nature [is] constituted through categories that carry normative force; if you want to 'develop' a wetland, call it a swamp; if you want to save a jungle, call it a rainforest" (1998, 5). Scott's central example of how state practices make nature legible is his brief history of the rise of German "scientific forestry" in tandem with the development of German state structures in the eighteenth and nineteenth centuries. As he argues, scientific forestry's "emergence cannot be understood outside the larger context of the centralized state-making initiatives of the period" (1998, 14).

If we apply Giddens's analysis of disembedding mechanisms to German scientific forestry, we can recognize in the efforts of German mathematicians to calculate a standardized tree—a *Normalbaum*—for each commercially useful species what Giddens has called the "creation of symbolic tokens." Efficient scientific forestry required large-scale, easily harvested forests; these were monocultures of evenly aged trees that "transformed the *Normalbaum* from abstraction to reality ... [and made the] German forest ... the archetype for imposing on disorderly nature the neatly arranged constructs of science" (Scott 1998, 15, citing Lowood 1991).

German scientific forestry became the model of forest management first for the rest of western Europe and North America, and then for much of the Third World. Of course, the development of the scientific forest radically altered the place-specific ecologies of old-growth forests from locally experienced, culturally meaningful environments to huge tracts of natural/national resources managed by cadres of experts in capital cities. Thus we see in scientific forestry the second of Giddens's two disembedding mechanisms, "the establishment of expert systems."

Giddens defines "expert systems" as "systems of technical accomplishment that organize large areas of the material and social environments in which we live today" (1990, 27). These are systems of abstracted, generalized expertise of which laypeople have only a weak understanding but in which they are required to place their trust as a condition of life in a complex modern state. Giddens gives examples of the architectural, aeronautical, and transport engineering systems that make so much of contemporary life possible yet remain largely opaque to the majority of those who spend so much of their lives in large buildings, on airplanes, or in automobiles on motorways.

Thus, as part of nation-state development, as in the case of German scientific forestry, we see the celebrated, romanticized Teutonic relation to forests reconfigured as a standardized product under the expert management of functionaries of the central state and then sold back to the people of Germany as a central image of their German-ness. The parallels with Foucault's studies of nineteenth-century practices of public health and population control (which Foucault grouped under the term "bio-power") are obvious. The *Normalbaum* concept was being developed at just the same period that French social scientists were constructing the *homme moyen* and the *famille normale moyenne* (the "average man" and the "normal average family") as basic units of urban planning (see Rabinow 1989, 1994). In much the way that the *homme moyen* was standardized and aggregated in the service of the state as citizen/worker/soldier, so Scott tells us, in German scientific forests,

rationally ordered arrangements of trees offered new possibilities for controlling nature. [cites Lowood 1991]

The tendency was toward regimentation, in the strict sense of the word. The trees were drawn up into serried, uniform ranks, as if to be measured, counted off, felled, and replaced by a new rank of lookalike conscripts. As an army, it was also designed hierarchically from above to fulfill a unique purpose and to be at the dispossession of a single commander. (1998, 15)

With the advantage of more than a century's hindsight, it is not surprising that the monocultures of scientific forestry soon fell vulnerable, rather like soldiers in the Great War, to the perils of standing in straight, even, homogenized ranks. This malady introduced a new word to the language, *Waldsterben* ("forest death") (Scott 1998, 20). German scientific foresters then scrambled to reintroduce some of the ecological diversity that had been extirpated with the old-growth forests. As Scott notes with some irony, this "restoration forestry" was undertaken "with mixed results to create a *virtual* ecology, while denying its chief sustaining condition: diversity" (1998, 21, italics in original).

German scientific forestry offers instructive lessons regarding the environmental implications of what Malkki calls the "arborescent metaphors" of nation building and for what she has called the "metaphysics of sedentarism" and its natural/national manifestations in maps and soils. Yet the disembedding mechanisms so evident in scientific forestry are also implicated in the scientific forest's ills. Forestry monoculture, designed to produce standardized lumber and other forest products for the German state and for transnational trade, not only led to *Waldsterben* but also leached specific forests of their local, historically sedimented meanings. Modern nation-state formation, although it frequently proclaims its advancement in the name of the national territory, its soil, mountains, forests, and waters, has in fact most often been carried out as an erasure of local or regional identity.

Further, the disembedding mechanisms that were developed and deployed in the service of the nation-state against regional cultures and ecologies have also had the effects of corroding the sovereignty of nation-states themselves. The classical territorial nation-state, with its cleanly bounded sovereignty over stable natural/national terrains, itself depends for legitimacy upon a steady-state metaphysics of nature and of political governance. Yet, the disembedding effects of modernity and late/post-/hypermodernity put in question the homeostasis of any natural or territorial entity.[8]

In this sense the "fluidity" of hydrological formations such as river basins and watersheds suggests that, if the dominant metaphors of territorial nationalism have been soil and trees, as Malkki persuasively argues, then the guiding metaphors of our era of flows will be hydrological or aquatic. Further, if water has been central to the historical organization of the modern nation-state form as territorial boundary and as infrastructure for transport and communication, it is fitting that water is now emblematic of the nation-state's erosion and/or transformation under the conditions of late—or postmodernity. The fluidity of water politics puts at issue the boundaries of any and all territorial entities and, taken together, the studies in this volume point toward new forms of sub-, supra-, trans- or paranational governance whose outlines may just be discernible as they precipitate out of this solution.

Notes

1. Even if the modern state defines itself as the sole legitimate repository of coercive force and of the means of internal and external violence (see Giddens 1985), any modern state that employs such force in other than exceptional circumstances risks abrogating the source of its own legitimacy—the consent of the governed—according to the terms of its own self-definition as a modern state. Although the modern rule of law is backed by threat of state violence, should violence come to be effectively substituted for the rule of law, the *Rechtstaat* has effectively delegitimated itself.

2. The notion of "sedimentation," or of historical interpretation as "desedimentation," is itself an appropriately aquatic term, borrowed from hermeneutics, for the effort to excavate the layers of sense laid down on the bed of any well-worn channel that guides sociocultural understanding; for a further explanation relevant to this volume's project, see, for example, Santos 1995, 457.

3. To take a single, representative example: Clive Ponting's widely read 1991 work, *A Green History of the World: The Environment and the Collapse of Great Civilizations*, takes no particular notice of the modern nation-state form in its 430 pages.

4. Renato Rosaldo has used the term "imperialist nostalgia" for this sentimental identification of the "vanishing native" with the degradation of the environment; it is, he says, "a particular kind of nostalgia ... where people mourn the passing of what they themselves have transformed.... When the so-called civilizing process destabilizes forms of life, the agents of change experience transformations of other cultures as personal losses." He describes how Euro-Americans "began to deify nature and its Native American inhabitants ... at the same time that [they] intensified their destruction of [North America's] human and natural environment" (Rosaldo 1989, 69–71). Benedict Anderson has referred to the appropriation of ethnic imagery to symbolize newly constructed nation-states as "logo-ization" (Anderson 1991, 178–185).

5. For one recent book that represents the global market as the historical culmination of the possibilities of human flourishing, see Stan Davis and Christopher Meyer, *Blur: The Speed of Change in the Connected Economy* (1998). The authors advocate that individuals "securitize" themselves, that is, issue stock in their own "human capital." The notion of "blur" is analogous to what I have been calling here "flows" (after Castells 1996), only even faster and more exuberant; this book's section headings include "The Blur of Desires" and "The Blur of Fulfillment."

6. The use of the Arabic term "jihad" by Barber and others to stand for any generic stereotype of militant particularism is, in my view, thoroughly regrettable. It propagates all-too-common stigmatizing Western imagery of Arabs and Muslims and relies upon an ignorance both of the long-standing universalist dimension of Islamic social thought and of the various late-nineteenth- and twentieth-century modernist reform movements in diverse Islamic societies, as well. See, for example, Malcolm Kerr, *Islamic Reform* (1996) and Maxime Rodinson, *Islam and Capitalism* (1974).

7. To quote Giddens once again:

it is not sufficient merely to invent new terms, like post-modernity and the rest. Instead, we have to look again at the nature of modernity itself which, for certain fairly specific reasons, has been poorly grasped in the social sciences hitherto. Rather than entering a period of post-modernity, we are moving into one in which the consequences of modernity are becoming more radicalized and universalized than before. Beyond modernity, I shall claim, we can perceive the contours of a new and different order, which is "post-modern"; but this is quite distinct from what is at the moment called by many "post-modernity." (1990, 2–3)

8. In addition, it is clear that classical territorial models of governance depend upon a steady-state vision of nature. Yet contemporary developments in the natural sciences suggest that homeostatic models of natural phenomena will always be inadequate and that the degree of their inadequacy is dependent simply on the time scale at issue.

Contemporary geological plate tectonics models of the dynamics of the Earth's surface—according to which land forms are constantly in flux—surely make nonsense of assertions of stable, naturally/nationally bounded territorial entities over the *longue duree*. Yet, even at more modest time scales of decades or centuries, it has become apparent that sociocultural, political, and economic expectations of stable territorial entities are tenuous.

It is not even necessary to make reference to earthquakes, volcanos, large meteorites, or other catastrophic cosmic or geo-seismic events. It has become clear that expectations about climate patterns on which immense agricultural economies (indeed, even the continued existence of the largest nation-states such as China and India depend) are not nearly as stable and predictable as rational state policies would require (see, e.g., Davis 1998).

From this perspective, the very notion of sustainable development, credibly reformist hope that it may be, is a hope founded on an impossibility, at least when assessed over the long term.

References

Anderson, Benedict. 1991. *Imagined Communities: Reflections on the Origins and Spread of Nationalism.* 2d rev. ed. New York: Verso.

Appadurai, Arjun. 1996. *Modernity at Large: Cultural Dimensions of Globalization*. Minneapolis: University of Minnesota Press.

Barber, Benjamin. 1995. *Jihad vs. McWorld*. New York: Random House.

Basch, Linda, Nina Glick Schiller, and Cristina Szanton Blanc. 1994. *Nations Unbound: Transnational Projects, Postcolonial Predicaments, and Deterritorialized Nation-States*. Amsterdam: Gordon and Breach.

Bernal, Victoria. 1997. "Colonial Moral Economy and the Discipline of Development: The Gezira Scheme and 'Modern' Sudan." *Cultural Anthropology* 12, no. 4: 447–479.

Boal, Iain. 1998. "Both Limbs and a Fork: Nature and Artifice on the West Coast." San Francisco: Yerba Buena Center for the Arts.

Burchell, Graham, Colin Gordon, and Peter Miller, eds. 1991. *The Foucault Effect: Studies in Governmentality with Two Lectures and an Interview with Michel Foucault*. Chicago: University of Chicago Press.

Castells, Manuel. 1996. *The Rise of the Network Society*. Cambridge, Mass: Blackwell Publishers.

Davis, Stanley M., and Christopher Meyer. 1998. *Blur: The Speed of Change in the Connected Economy*. Reading, Mass.: Addison-Wesley.

Davis, Mike. 1998. *The Ecology of Fear*. New York: Metropolitan Books.

Ferguson, James. 1990. *The Anti-politics Machine: "Development," Depoliticization and Bureaucratic Power in Lesotho*. New York: Cambridge University Press.

Foucault, Michel. 1972. *The Archaeology of Knowledge*. 1st Amer. ed. New York, Pantheon Books.

Foucault, Michel. 1980. "The Eye of Power." In Colin Gordon (ed.), *Power/Knowledge: Selected Interviews and Other Writings, 1972–1977*. New York: Pantheon.

Fukuyama, Francis. 1992. *The End of History and the Last Man*. New York: Free Press.

Gellner, Ernest. 1983. *Nations and Nationalism*. Oxford: Blackwell.

Giddens, Anthony. 1985. *The Nation-State and Violence*. Berkeley: University of California Press.

Giddens, Anthony. 1990. *The Consequences of Modernity*. Stanford, Calif.: Stanford University Press.

Habermas, Jürgen. 1987. *The Philosophical Discourse of Modernity*. Cambridge, Mass.: Polity Press.

Hannerz, Ulf. 1996. *Transnational Connections: Culture, People, Places*. New York: Routledge.

"Harper's Index." 1997. *Harper's*. April. pp. 13, 87.

Ignatieff, Michael. 1993. *Of Blood and Belonging: Journeys into the New Nationalism*. New York: Farrar, Straus and Giroux.

Jefferson, Thomas. 1800. *Jefferson's Notes, on the State of Virginia*. In series *Early American imprints*, first series, no. 37702. Baltimore: W. Pechin.

Judt, Tony. 1994. "The New Old Nationalism." *New York Review of Books* (May 26): 44.

Kerr, Malcolm H. 1966. *Islamic Reform: The Political and Legal Theories of Muhammad 'Abduh and Rashid Rida*. Berkeley and Los Angeles: University of California Press.

Lash, Scott, and John Urry. 1994. *Economies of Signs and Space*. London: Sage Publications.

Latour, Bruno. 1993. *We Have Never Been Modern*, trans. Catherine Porter. Cambridge: Harvard University Press.

Lovejoy, Arthur O. 1948. *The Great Chain of Being: A Study of the History of An Idea*. The William James lectures Delivered at Harvard University, 1933. Cambridge: Harvard University Press.

Lowood, Henry. 1991. *Patriotism, Profit, and the Promotion of Science in the German Enlightenment: The Economic and Scientific Societies, 1760–1815*. New York: Garland.

Malkki, Liisa. 1992. "*National Geographic*: The Rooting of Peoples and the Territorialization of National Identity among Scholars and Refugees." *Cultural Anthropology* 7, no. 1 (February): 24–43.

Montesquieu, Charles de Secondat. 1834. *De l'esprit des lois*. Paris: Didot.

Ong, Aihwa. 1999. *Flexible Citizenship: The Cultural Logics of Transnationality*. Durham, N.C.: Duke University Press.

Perry, Richard Warren. 1995. "The Logic of the Modern Nation-State and the Legal Construction of Native American Tribal Identity." *Indiana Law Review* 28, no. 3: 547–574.

Ponting, Clive. 1991. *A Green History of the World: The Environment and the Collapse of Great Civilizations*. New York: Penguin Books.

Rabinow, Paul. 1989. *French Modern: Norms and Forms of the Social Environment*. Cambridge: MIT Press.

Rabinow, Paul. 1994. "On the Archeology of Late Modernity." In Roger Friedland and Deidre Boden (eds.), *NowHere: Space, Time, and Modernity*. Berkeley and Los Argeles: University of California Press, pp. 402–418.

Robertson, Roland. 1995. "Glocalization: Time-Space and Homogeneity-Heterogeneity." In M. Featherstone, S. Lash, and R. Robertson (eds.), *Global Modernities*. London: Sage Publications.

Rodinson, Maxime. 1974. *Islam and Capitalism*, trans. Brian Pearce. 1st Amer. ed. New York: Pantheon Books.

Rosaldo, Renato. 1989. *Culture and Truth: The Remaking of Social Analysis.* Boston: Beacon Press.

Santos, Bonaventura de Sousa. 1995. *Toward a New Common Sense: Law, Science, and Politics in the Paradigmatic Transition.* New York: Routledge.

Scott, James C. 1998. *Seeing like a State: How Certain Schemes to Improve the Human Condition Have Failed.* New Haven: Yale University Press.

Ward, Colin. 1997. *Reflected in Water: A Crisis of Social Responsibility.* London: Cassell.

White, Lynn Jr. 1966. "The Historical Roots of Our Ecological Crisis." In Theodore D. Goldfarb (ed.), *Sources: Notable Selections in Environmental Studies.* Guilford, Conn.: Brown and Benchmark.

12

Lessons from the Spaces of Unbound Water for Research and Governance in a Glocalized World

Richard Perry, Joachim Blatter, and Helen Ingram

This book is about water—how it is understood, how it is governed—in the world at the beginning of the third millennium as the territorial bond between the natural and the national is fraying. The research strategy that characterizes this book is multiple, methodologically antireductionist, and conceptually heterodox, as it calls into question received understandings and explores paradoxes on the basis of empirical studies. Thick description, or digging deep into historical and local knowledge, grounds the efforts in the studies represented here to infer a conceptual framework that can connect various localities and interpret local narratives in light of wider concerns. To fit the case studies into a broader framework in this chapter obliges us to reflect critically on chapters 1 and 2 and to recognize that not all the approaches we recommend are equally well suited to particular situations and that the transformations we have observed in progress may be limited in their scope.

Our stance toward the case studies presented in this volume is reflective and critical as we seek to draw inferences and parallels that the authors themselves may not have emphasized. In the effort to draw theoretical lessons from disparate cases we must take the risk of proposing interpretations that the authors themselves, embedded as they were in their particular narratives, may not have viewed as most salient. The great advantage offered by an overview of eight local cases is the possibility of drawing connections and contrasts on a more general level. The purpose of interpreting diverse cases in light of a general framework is to draw lessons for research and governance in a glocalized world. We close this book with a call for diversity and variety of modes and institutions of governance. In light of evidence in our case studies, we conclude that effective governance structures and arrangements must reflect specific contexts.

Interpreting the Case Studies: Dialectical Development, Contradictions, and Coexistence of Premodern, Modern, and Postmodern Meanings of Water

In the review that follows, we find evidence offered by the various case studies in the book that both supports and contradicts the contentions of the introductory chapters. Across the specific contexts, the postmodern heterodoxy of research approaches and the shifting meanings of water yield interesting and important insights. There is no uniform evolutionary schema either in the meanings of water or in the appropriate research strategies.

Modern state-based approaches continue to be of use in explaining transboundary relations, both in economically advanced and culturally compatible regions of the world and in areas of ethnic and national conflict. Moreover, evidence that quite different meanings coexist in a number of localities undermines any claim that meanings of water are evolving in a unilinear trajectory from pre- to postmodern. We do, however, find the broad range of meanings that we anticipated in chapters 1 and 2.

María Rosa García-Acevedo's historical "thick description" in chapter 3 of the flows of water, population, capital, and development projects across the U.S.-Mexico border over the past century provides a wonderful lesson in the layering of meanings of water over time. She describes different meanings of water that emerged through a historical dialectic. In the precolonial period, the indigenous population of the area was embedded in a riparian cultural ecology that, in the framework sketched in chapter 1, perceived water as a natural given, a gift of nature. The close relationship of the indigenous people to water is expressed in their name: Cocopa—"those who live in the river." The seasonal climatic cycles of the Colorado River shaped the Cocopa culture, and the Cocopa did not themselves intervene in the Colorado River systems on any large scale.

Nation-state governance was as alien as water engineering to the Cocopa. The fixing of the U.S.-Mexico border by the Treaty of

Guadalupe Hidalgo in 1848 had little impact on them. The national border was relatively porous in the nineteenth century, and the inflows of capital and people, far more than the exercise of state power, were the primary agent of change in the area. The logic of economic expansion that accompanied the American and Mexican settlers who arrived in the region, along with their organizations and technologies, sharply affected the indigenous culture. The "spaces of capitalist flows" were not bounded by the borders of the two nation-states.[1]

Capitalist development and ideology in the lower Colorado River basin brought with it new meanings of water as property and product through which water could be disconnected from its river bed and its traditional role in indigenous society. Water flowed northward toward growing populations and accumulating wealth, little constrained by borders. Only later did the logic of nation-state interests and national boundaries play an important role in governing the region.

Catastrophe (floods) and ideology (land reform) invited an enlarged federal presence as water became a security issue and a tool for redistribution of wealth. To legitimize redistribution of water between groups, the nation-states needed a concept of a common identity: the nation as an "imagined community" (Anderson 1991) that transcended ethnic, cultural, and place-based differences.

Once the U.S. and Mexican central governments began to intervene in the regional ecology of water, the nature of the border and political relations between the Mexicali and Imperial Valleys changed dramatically. The two governments undertook to divide Colorado River water between them and to enforce control of the border between the two countries. This effort continues to the present day even though the primary focus of state exertion has changed from regulating the flow of water to controlling the flow of people.

Most recently the dominant meaning of water in the Mexicali and Imperial Valleys has again been transformed, now to water as commodity. A more porous border has returned with this altered meaning. The U.S.-Mexico border is now easily permeable for capital and goods but much less so for people. García-Acevedo's study in chapter 3 supports a more general characterization of the difference between the course of transnational integration in Europe and North America under the European Union and the North American Free Trade Agreement. Whereas the (western) European leitmotiv for continental integration is a holistic, encompassing community, a "common European home," in North America, the dominant model is an economic alliance and little more (Blatter 1998).

García-Acevedo observes that, for now, an enthusiasm for the market commodification and price allocation of water dominates policy discourse in both the United States and Mexico. As she astutely notes, however, this profoundly disembedded approach to water is likely to bring conflicts within and between the two countries along several fault lines: rural versus urban/suburban communities, indigenous peoples versus majority populations, and cross-border coalitions of advocates of economic development versus environmentalists who embrace a more conservationist ecological vision. These cleavages may generate an explosive mixture of value conflicts both within and between the two countries.

García-Acevedo's case study shows the rich insight into the historically shifting meanings of water that can be gleaned through historical inquiry. Only such in-depth, locally grounded understanding can teach us who the dominant actors are and whether the conflict is a "war" (invasion of settlers), a "game" (distributional conflicts between groups), or a "value conflict" (because of the noncommensurable normative preferences of the involved groups). Only on the basis of such a nuanced local understanding can the formal models and quantitative analyses advanced by modern water scholars be useful.

In contrast, Joachim Blatter's study in chapter 4 of the growth of networks of subnational coalitions and institutions in the contemporary regulation of leisure boat pollution of Lake Constance shows us that very different meanings of water are coming to the fore in some contexts. Blatter's history of transborder environmental regulatory developments is distinctive in several ways. First, the historical regional identity of the riparian interests surrounding Lake Constance has carried over from former eras, and Lake Constance remains subject to a "condominium" regime that Blatter describes as a "pre-Westphalian curiosity in international law." This is very different from the usual mutually exclusive territorial division of bodies of water between contiguous states.

Second, this enduring, regionally identified regime persists in the context of a shared cultural linguistic heritage among the riparian nation-state regions: Germany (successor state to several earlier German-speaking kingdoms, grand duchies, and principalities), Austria (successor state to the diverse Austro-Hungarian Empire), and the German-speaking Swiss cantons. All of these are historically Germanophone and distinctively Alpine in cultural identity. Such close transboundary commonalities cannot be assumed in many other parts of the world.

Third, despite earlier political conflicts dating back to the medieval era, Lake Constance has in modern times been the object of very little political-security conflict. In the environmental-regulatory questions that Blatter discusses, it is clear that the conflicts involve neither central state security issues, primary nation-state issues, nor even regional economic interests, but rather tourism and leisure time activities. Tourism and recreation, although they generate enormous financial returns, are the quintessential postmodern, poststatal economic sectors. Tourism-leisure consumption is the ideal type of postindustrial service economy in global flows of commerce.

In his study of functionally and sectorally defined networks of transboundary institutions, Blatter offers what may be a best-case scenario for a regional environmental regulatory regime. The regulatory regime governing Lake Constance has historical and cultural roots in a surviving premodern condominium, a regime that has been revivified as a "postmodern" example of what Blatter calls the "polity ideal of a 'Euroregion' inspired by the European unification process." Blatter has shown us a case of transborder environmental governance in which outcomes depend less on nation-state sovereignty than upon sub-, supra-, or parastate initiatives undertaken within the framework of a well-developed shared regional environmental discourse. This case study closely fits the framework in chapters 1 and 2, notwithstanding questions about its generalizability.

The second case study that demonstrates a post-nation-state, postmodern approach to water is Suzanne Levesque's analysis in chapter 5 of the Yellowstone to Yukon Conservation Initiative. Levesque's study shows a border that has less separated than simply demarcated the territorial jurisdiction of two wealthy Western countries with a long history of shared culture, language, and peaceful relations. The supranational connections that enable the construction of cross-boundary networks are even more obvious here than in the Lake Constance case. The Y2Y territorial initiative transcends nation-state boundaries by disregarding them, and its membership defines itself by shared values and transnational circuits of communication. Its membership is connected by a non-territorial bond, the Internet, rather than through modern linkages such as common citizenship or premodern ties such as kinship. The Y2Y network is also postmodern in that it is an issue-specific community that aims to influence many governments at multiple levels rather than to establish an autonomous multipurpose governance unit such as the "Sovereign State of Y2Y."

A seemingly contradictory aspect of this case study is the fact that the environmental community is interconnected by "postmodern" flows (e.g., information, ideas, and visions), but uses "premodern" natural flows (e.g., watersheds and animal migration paths) as focal points for identity building. The language Y2Y members use suggests that their beliefs are "essentialist." The activists' language refers to "love of the land" and "sense of place," concepts that would seem to frame the connection as one of objective need and natural imperative. The notion that humans need some anchorage or belief that is beyond rational human control is thus at once premodern and postmodern.

Like many environmentalists, the Y2Y participants use the word "place" in an organic, holistic sense to signal the "connectedness" of human existence to the local natural and cultural environment. Our reading of the case, however, is that Y2Y members experience their "sense of place" within the "spaces of flows" and are therefore ultimately postmodern (Castells 1996, 1989). The initiative is idea driven, and its notions of boundaries are only roughly delineated by its assumptions about wildlife corridors and migratory routes. Moreover, there is no "natural imperative" that correctly dictates a precise boundary for governance. After all, the choice of megafauna habitats as a biologically based boundary determiner, rather than other, equally viable ecological markers such as watersheds, is itself a value claim. Even though its members see it as a natural imperative, the Y2Y is a socially constructed conception.

Y2Y is at least as much an idea-driven as an interest-driven concept. A modern scholar might wish to write into the case study a portrayal of an activist organization that purposefully manipulates images and symbols to get support for controlling or governing the environment in its own interest. The rich detail of the case that Levesque provides suggests otherwise. The following quotation illustrates the role of emotional commitments in the foundation of the network: "There is no doubt in my mind that the people in this room love the land as fiercely as a mother grizzly loves her cubs. And are prepared to fight for it just as fiercely as a mother grizzly."

The authors of chapters 6 and 7, Kathleen Sullivan and Pamela Doughman, both employ discourse analysis in their studies. They assume from the outset that discursive practices have an impact on transboundary water policies. Nevertheless, both studies find that the actors in transboundary water management have not embraced postmodern meanings of water. In neither case presented in the two chapters is transborder discourse dominated by concepts such ecosystem integrity or environmental sustainability; such ideas remain peripheral. Instead, the actors observe the terminology of modern nation-state interests, such as "natural resources management" and "the fisheries war."

Sullivan finds contradictions in the discursive sphere of salmon fisheries in the Pacific Northwest. She has given us an illuminating study of the "salmon wars," the ongoing contemporary conflict between Canadian and U.S. interests over salmon fisheries and their management. As Sullivan examines both expert and public media discourses surrounding the salmon wars, she interweaves two quite different, even opposing, approaches to discourse analysis. On the one hand, she engages the modernist Frankfurt School critical theory approach to public media as it has been influentially formulated by the leading German socio-legal theorist, Jürgen Habermas. On the other hand, Sullivan seeks to integrate into her analysis the work of French thinker Michel Foucault. These two approaches to discourse analysis are quite sharply opposed. Whereas Habermas points to the potential of discourse in the public sphere for arriving at shared "commonsense" solutions,[2] Foucault is much more skeptical as he seeks to bring to view the power relations underlying all discourse.[3] Sullivan finds support in her study both for the democratizing possibilities of transnational media flows of information and for the play of power in discursive fields as well.

The regional struggle over wild salmon transcends national boundaries and illustrates the glocalization process taking place in the contemporary world. At the same time, nation-states and nationalism continue
to have a strong role in determining fisheries policy. International law
instruments, agreed upon through binational negotiations, established
the themes of equity and conservation that have become weapons in the
highly visible media confrontation between Canadian fishers and U.S.
officials. The equity theme became the basis for grassroots action in
which access to international media outlets was critical. The conservation theme, however, was colonized by the technocratic rationality implicit in complex scientific models built on value assumptions. The
resulting debate, conducted in the language of experts, was put beyond
the reach of the fishers whose way of life was at issue. Transnational
mass media, while far from providing equal access, at least allowed a
forum for local expression. In this case, local interests in Canada and the
United States are sharply at odds. Therefore, this case study further cau-
tions us not to overgeneralize about the postmodern networks emerging
on the U.S.-Canada border. The sort of conservationist alliance built
upon common belief structures described in chapter 5 clearly does not
extend to salmon harvesting.

Doughman observes that the NADBank and the BECC share a discursive framework in the rhetoric of "sustainable development." She notes
that "sustainable development" is a relatively recent arrival in the discourse of "global civil society." From a constructivist, postmodern perspective, the term is an innovation in development discourse that has
evolved in response to the widespread critiques of the destructive effects
of traditional development programs, especially of the "resource extraction" model. The innovative element of this phrase is its holistic, encompassing notion of sustainability, which bridges several disciplinary
boundaries. Since the environment is no respecter of political jurisdictions, sustainability transcends territorial borders. But even more important, sustainability bridges the boundaries between economic, social, and
political spheres, since equity, economic well-being, and open participation are its conditions of possibility.

Doughman's critical stance and methodology reveal that "sustainable
development" can also be seen as a phrase that may not only obscure the

divergent underlying interests of the actors but also contribute to miscommunication. Doughman's data nicely show that there exist wide divergences of perception among people who are involved with or affected by water development projects in the U.S.-Mexico border region. Some of Doughman's informants are enthusiastic supporters of sustainable development, seeing it as a new and hopeful model for public participation and deliberative democratic governance. Others are suspicious of the premises of sustainable development, seeing it as just a new spin on old, familiar practices of exploitation of nature and of subordinating populations.

The next two case studies presented in the book analyze contexts far removed from the relatively peaceful, culturally continuous transborder areas of Europe and North America. They remind us that although many social scientists who work in these areas may be observing emergent "postnational constellations" (Habermas 1998), in other parts of the world the current direction of change is toward the collapse of transnational empires such as the Soviet Union and the painful, conflict-ridden consolidation of new states. The case studies by Paula Garb and John Whiteley and Joe DiMento in chapters 8 and 9 show that, although there are hints of new meanings of water, modern logics of strategic state interests prevail.

The study of the Inguri River hydroelectric facility by Garb and Whiteley shows us a context in which claims to nation-state status are being asserted from the shards left over after the collapse of a former overarching empire. Here in the Caucasus, out of the rubble of the former Russian/Soviet empire, both Georgia and Abkhazia have sought internationally recognized nation-state status. The former Soviet "constituent republic" of Georgia has, as of 1999, been internationally recognized and Abkhazia, formerly an "autonomous republic" within Georgia, is now asserting its own claim to national self-determination, although it has not received international recognition at the time of this writing.

These two polities are currently in an uneasy state of armed stalemate. The fact that the Inguri River marks the military frontline makes it more than likely that water would be seen as a security issue in the area. The puzzling and hopeful finding of Garb and Whiteley's study is that the essentialistic meaning of water as security issue does not necessarily result in an uncooperative confrontation, as the framework developed in the introductory chapters would suggest. How can we explain why, in such a premodern "realist" context, neither of the two nations at war in this situation has used the power plant as a weapon to destroy the other side? Such aggressive action certainly would seem likely given that both sides claim territory as their "historical homelands" and a premodern tradition of blood revenge among some of the participants has been revived.

Several explanations are possible. The modern, rational explanation is that the leaders of the two nations care more about their economic interests than about the ethnic strife that would have Georgia consolidate itself by destroying Abkhazia, even if that meant loss of its electric power for a significant period. Therefore, water as a security issue for building the nation-state may be more connected to a utilitarian logic than to the "relative gains" logic of a premodern meaning.

The second explanation for cooperation draws upon postmodern as well as modern elements. It suggests that the European Bank and the United Nations are using their legal and economic influence to foster cooperation among the two warring parties in the management of the complex. Insofar as this explanation engages supranational actors and their networks, it engages postmodern theory.

The third explanation relies on the shared beliefs among members of a professional "epistemic community" who keep the power plant in operation. The epistemic community in this case study simultaneously possesses features we would categorize as postmodern (it is a transnational community), modern (the predominant focal point for this community of engineers is the meaning of water as a modern technological output), and premodern (the bounds of this community are determined not solely by their common technocratic view of water but also by traditional sociocultural ties: trust between the top engineers, banquets, toasts, and other ceremonies).

In the spirit of critical reflection, we conclude from this case study that in many parts of the world "modernist" approaches toward transborder water management remain quite useful in avoiding damaging conflict. At the same time, although open conflict has been avoided through enlightened leadership and technical cooperation, serious environmental externalities related to the generation of hydropower have yet to be addressed. Putting such broader environmental issues on the agenda for negotiation would require the construction of the kind of trans-boundary environmental networks about which Blatter and Levesque write in their chapters.

Exactly this hope that more environmentally sensitive actions can take place in a similar conflictual context undergirds chapter 9, Joseph DiMento's study of the Black Sea. The case DiMento presents illustrates a diligent modern legal thinker at work. He believes that there has to be a "solution" to save the Black Sea, even though he reports that his efforts to find any possibility of a solution have been largely without fruit. New states where environmental ministeries are weak and the political leaders have other priorities offer little hope of a solution. There is only scant promise of such from a tenuous civil society. The nongovernmental organizations in operation in the area survive on shaky foundations, propped up by funding from the West. There is more potential in modern and postmodern Western instruments of international legal regimes and transnational consortiums of scientists, but DiMento concedes that the first head of the Black Sea Environmental Program gave up in frustration. It is telling that the chapter closes with the vision of a future in which shared environmental ethics come to bridge territorial and sectoral cleavages.

Having classified Joe DiMento's approach as modern, we recognize that his endeavor, with which we have great sympathy, is not one many positivists, who would be satisfied with a realistic appraisal, would endorse. Accurate description and analysis are not enough. Normative goals, explicitly laid down and open for criticism, are also legitimate scholarly work and, in the long run, more useful. Identifying instruments for better environmental cooperation in the Black Sea region is a laudable exercise, despite the difficulty inherent in their implementation. Planting and nurturing the seeds for environmentally sound development may well be more helpful than a realistic analysis that could dampen whatever possibilities exist.

David McDermott Hughes's discussion in chapter 10 of water and boundary disputes in the Chimanimani highlands of Zimbabwe brings us back to the themes of indigenous people's relations to water that García-

Acevedo touched on in the book's first case study. In both Hughes's African case study and García-Acevedo's study from North America, water is an inextricable part of the natural environment rather than an economic or national resource under human control. For the Cocopa people who live on arid lands in the Colorado River delta, water determined all aspects of life. In contrast, in the humid uplands of Zimbabwe, the linear aspect of natural water is important. The lines natural watercourses trace on the land are meaningful referents indigenous peoples employ to mark the lived terrains of their homelands.

This was hardly the case with the colonial rulers in what is now Zimbabwe nor with the bureaucratic postcolonial governments that succeeded them. Instead, these governments have adopted characteristically modern approaches to boundaries, using survey instruments to draw straight lines that ignore indigenous knowledge of patterns of habitation and harvesting as well as the practical utility of streambeds as boundary markers in mountainous terrain. The clear-cut apportioning of territory through modern surveys is not conceptually different from the appropriation of volumetric water rights and the diversion of water as product and property in the lower Colorado River about which García-Acevedo writes. Both modern practices are impositions of government control not only over nature but also over the land's first human inhabitants.

The in-depth historical method that Hughes employs allows us to see boundary disputes as they have played out over time in struggles over definitions of nature. Hughes is an anthropologist who tells the story as a participant-observer. He explicitly seeks to portray the perspective of the indigenous community. Our reading of the case puts more stress on the emergence of "postmodern" meanings of water than Hughes does. One of the boundary disputes that Hughes describes concerns an estate that has been purchased by a new owner as a butterfly sanctuary, a postmodern land use. The valley bounded by the headwaters must be set aside, according to the owner, to preserve the butterflies.

The importance of water as an integral part of wildlife habitat is a postmodern idea introduced by Levesque in chapter 5. It is not at all surprising that a wildlife enthusiast is far more willing than previous estate owners to adjust peripheral boundary lines and abandon the straight line that marks his legal tenure. As we have already suggested in our review of chapter 5, such phenomena as wildlife corridors, bio-regions, and ecologically sustainable habitats are socially constructed in such a way that concern for modern notions of private property rights is subordinated to an underlying core value of preserving natural ecological formations.

A new array of actors is also visible in Hughes's account of the boundary dispute over the Rusitu Botanical Reserve. Here, the bureaucratic Department of National Parks chose to assert straight-line boundaries that were neat in the abstract but defied natural boundaries by plunging lines over cliffs and ignoring historic land occupancy patterns that conformed to stream beds. In defense of their land rights, the indigenous population, in partnership with an international nongovernmental organization (INGO) devoted to community-based natural resource development, resorted to the unprecedented strategy of making their own countermap of the terrain, one that depicted traditional streambed boundaries. This partnership of international organizations with the local peoples nicely exemplifies the glocal governance patterns that we proposed in chapter 1 as an emerging paradigm. Moreover, the INGO's previous interest in nudging a committee of locally based prominant people into ecotourism, a postmodern leisure activity, links this case firmly to the broader theoretical framework of this book. The studies collected here suggest the possibility of models in which water and power, nature and culture, are globally and locally reconstituted and are recognized as embedded in what are irreducibly social ecologies. These studies exemplify the myriad possibilities of local-global dynamics of flows in which water governance is paradigmatic of "glocalization," where water and power are inextricably interwoven.

Lessons for Water Research and Governance in a Glocalized World

Environmental writer Kirkpatrick Sale, in an editorial sardonically titled "Liquid Asset," advised readers to "mark the date: March 21, 1998 ... [for o]n that day, water—yes, water, the H_2O of oceans, rivers, lakes and rains—officially became a corporate commodity" (Sale 1998). Sale is referring to the declarations of a UN conference on water that, as Reuters reported, specifies that water henceforth "should be paid for as a commodity rather than be treated as an essential staple to be provided free of cost" (Sale 1998, 58). French Prime Minister Lionel Jospin congratulated the delegates for adopting a "prudent" market-oriented approach and for having renounced the old, outmoded idea, "which held on for too long, that water could only be free because it fell from the heavens" (Sale 1998, 7, 58). In a further step, French President Jacques Chirac then proposed a global water market system managed by private (i.e., non- or supragovernmental) corporations. The ascendency of such a narrow and limited conception of water and water governance is precisely what this book has set out to challenge. Briefly stated, to better understand water politics in a glocalized world, we need to unbound the narrow meanings of water and to restore or re-embed water in its natural and cultural contexts.

It is crucial to the formation of a more complete understanding of water policy to recognize and overcome the limits of the disciplines that dominate the modern understanding of water: engineering, law, and economics. These disciplines share a common approach to nature, what Heidegger (1977) termed the "technological understanding of Being." This implies that water, removed from its context through engineering technologies, becomes a resource for humans to manage and distribute for utilitarian ends (cf. Giddens' "expert systems"; Giddens 1990 and chapter 11). Next, water is apportioned into "volumetrically" quantified, fungible units of economic market exchange and circulation (cf. Giddens' "symbolic tokens"; Giddens 1990 and chapter 11). Then these commodified units of exchange are allocated and distributed through legal regimes of property rights to potentially worldwide or global networks. Many contemporary thinkers join Chirac in his wish to see global markets in which transnational private actors can engage in water transactions. For us, this is neither an empirically accurate nor normatively defensible model for water resources governance in the new millennium.

This book has engaged multiple disciplines and emerging approaches to reveal very different meanings for water and visions of the glocalized world. We argue that a narrow focus on rational actors and their utilitarian preferences explains a decreasingly important and less interesting aspect of human interaction with the environment. Cultural traditions and historic understandings as well as the emergence of new transnational communities are asserting different identities based on ideas quite distinct from material wants. The creation and constitution of communities and political actors must be taken into account in analyzing transboundary water politics.

As noted in chapters 1 and 2, new models are needed for understanding the plural meanings of water and more generally for understanding the natural world as a whole. We have seen that modern technocratic, scientistic forms of knowledge and governmental practice are themselves inextricable from the long-standing civilizational narratives of nation-state formation, development, and modernization. This is the sort of explanatory framework—reason, science, governance, progress—that Jean-François Lyotard has famously called a *meta-recit* (i.e., a "meta-narrative" or a "grand-narrative"; see Lyotard 1985 and Habermas 1987). From this perspective we can see that recent controversies about "globalization" and "postmodernity" concern precisely the question of whether these phenomena mark the ultimate fulfillment of the civilizational story, or whether they mark a turning in a new direction, one that may afford new clearings, new vistas for thought and practice.

As Giddens has summarized the debates over the shift from one historic socioeconomic framework "based upon the manufacture of material goods to one concerned more centrally with information," the controversies have "focused largely on issues of philosophy and epistemology." Giddens cites Lyotard's view that

post-modernity refers to a shift away from attempts to ground epistemology and from faith in humanly engineered progress. The condition of post-modernity is distinguished by an evaporating of the "grand narrative"—the overarching "story line" by means of which we are placed in history as beings having a definite past and a predictable future. The post-modern outlook sees a plurality of heterogenous claims to knowledge, in which science does not have a privileged place. (Giddens 1990, 2)

If it is indeed true that we are entering an era distinguished by an epistemological paradigm shift (see, e.g., Santos 1995), then the "plurality of heterogenous claims to knowledge" will likewise require a plurality of approaches to questions of the nature of nature and of governance. A potentially infinite range of intellectual modes of inquiry must be considered. To do justice to the facts as we encounter them "on the ground" will require something like what Clifford Geertz has famously termed "thick description," which seeks to elucidate the myriad forms of "local knowledge" that constitute humankind's encounters with the natural (Geertz 1983). Such an inquiry would of course include the sorts of history of thickly sedimented local meanings that Geertz advocated, as well as the interpretive methods of discourse analysis that Geertz—along with Foucault, Goffman, and others—has theorized and applied since the 1960s. A study of the concrete modes of governance of water and other natural phenomena must also attend to the properties of emergent networks of institutional governance that come into being in so many transborder contexts. All of these approaches are both holistic and locally grounded, even when the localities studied are globally dispersed. We can envision them, taken as an open-ended ensemble as truly social ecology: a socially grounded, nondogmatic, ecological approach that reflexively considers its own "ecology of knowledge" (Bateson 1972).

In this volume we have sought to build on "locally grounded" case studies in an effort to imagine a broader conceptual framework that might make possible shared understandings across localities and among researchers working in different contexts. Such a framework would enable local knowledge from one context to be received as something more than "noise" in another. The goal of the schema, provided in chapters 1 and 2, that represents ontological transformations over time in the layering of changing meanings of water is something other than the methodological reductionism characteristic of modern metatheories. Rather, our endeavor is to identify other orders of sense and to make understandable the differences among them. Where modern approaches seek to eliminate uncertainty and ambiguity, we aim to reveal contradictions and paradoxes. The only possibility of integration starts with reflection upon and acceptance of difference and particularity and proceeds by tentatively drawing pragmatic, infinitely revisable generalizations from across the diversity of local circumstances.

Among the diversities and complexities this book embraces is the co-occurrence of premodern, modern, and postmodern meanings of water. Postmodern conceptions must be thought of not as supplanting but as supplementing previous meanings. As the case studies presented in the book demonstrate, modern meanings of water continue to resonate strongly in many parts of the world. Moreover, under some circumstances, agreement on modern meanings facilitates cross-boundary understandings. The case studies also confirm the presence of postmodern meanings of water emergent in vanguard situations, possibly signaling the direction of future change. For water researchers to be able to engage in useful comparative analysis, they must look carefully at similarities and differences in motivations and values of actors. Although a good deal can be learned by carrying insights from one context to another when meanings are parallel, imposing explanations that fit well in one circumstance into another in which meanings are fundamentally different leads to misinterpretation and damaging prescription.

Another seeming paradox that this framework brings to view is the continuing importance of the nation-state at the same time that cross-border communities and new nonstatist modes of governance are gaining ground. Whereas conventional modern research methods may elucidate nation-state behavior, other methods are more appropriate to the understanding of alliances and advocacy coalitions built on shared belief systems. Moreover, particular attention needs to be paid to the ways in which new kinds of alliances are bringing different pressures upon the nation-state, through domestic politics and behavior in the international arena. Intermestic politics is an area of growing importance, and the subdisciplinary lines between foreign and domestic policy poorly serve contemporary research needs.

Transboundary water governance is not likely to become less complex, or even less conflict-ridden, in the emerging world of multiple transboundary ties and relationships. This book recommends no specific institutional formula or technical design for apportioning water across boundaries. Instead, our aim has been to expand the possibilities of new modes of governance beyond those conventionally proposed. Cross-border advocacy networks, public discourse, and the emergence of cross-border regional identities cannot be ignored in approaches to managing transboundary flows.

In efforts to minimize conflict, the involvement of new actors and new institutions may have great utility. They "unbundle" the boundaries represented in conflicts over natural resources (Elkins 1995). When the border between nation-states is no longer the unique line of division, the danger of deadlock is reduced. A world in which a diversity of social actors can express a fuller variety of meanings of water allows greater possibility for cooperative deliberation. Nevertheless, these developments may turn out to be no more than epiphenomenal if no institutional structures exist capable of implementing sustainable development or ecosystem management programs. Further, cross-boundary discourses and coalitions founded on shared beliefs in postmodern meanings may not, in fact, facilitate cooperation. They may instead result in confrontational constellations unable to find middle ground between nonnegotiable belief systems, especially when there are neither cross-cutting issues nor overlapping memberships that mitigate conflict. Gridlock may be an even less satisfactory outcome than the bargains struck between the bureaucratic agents of nation-states as they are typically brought together in international commissions under the modern approaches to transboundary resource management.

The world of transboundary water governance that we envision would exhibit the same kind of multiply differentiated institutional structures and relationships we see in domestic systems. A broad variety of modes of governance would coexist in robust civil societies that would undertake the responsibilities for water governance in many parts of the world. International water market systems might come to play an important, though limited, role. In other areas, transboundary water relationships would be governed by traditional nation-state actors maintaining sovereign power within international legal regimes. Differences would exist not only across regions, but also among different understandings of water issues as they are variously framed and defined as leisure activities, transportation, hydroelectric power, domestic water supply, and all the other issues on which water touches.

If transboundary water governance is to be improved in the future, we believe greater openness and participation must accompany the multiplicity of evolving institutional structures. Transboundary water decision making can no longer be dominated by a narrow class of individuals with political power and technical expertise. Water must be unbound from the narrow strictures within which it has been considered in the past and revivified as part of a more inclusive natural and human environment.

Notes

1. As Andrew Ross argues, environmentalism

is now being ... assimilated into corporate logic by the post-Rio move to create a world environmental market in the form of free-market solutions to the problem of absorbing, distributing, and exploiting environmental costs. Between the World Bank–administered Global Environmental Facility, set up as a green fund for development, and the work of the Rio Summit's Business Council on Sustainable Development, a global market has emerged to rationalize everything from debt-for-nature swaps to the expedient bartering of individual states' environmental regulations, all in the interests of smoothing the flow of international finance. Far from self-regulating, this free market is governed by a cost-benefit budgetary model, but nonetheless presented as the best that economists have to offer in the way of a 'natural' solution to the crisis of nature. (1994, 127)

For a more extensive critique along these lines, see McAfee (forthcoming).

2. This idealized public sphere posited by Habermas is characterized by a discursive ethics that guides nonhierarchical, nondominative forms of communicative action oriented toward achieving shared understanding and ultimately a democratically legitimate consensus on norms, facts, and values (Habermas 1996).

3. For Foucault (1995), such a Habermasian vision of discourse and knowledge that escapes the play of power and human interest represents at best a naively ahistorical apology for modern Western political forms and is arguably complicit in the myriad forms of coercion in modern state governmental practices. In Foucault's view, knowledge and power are inextricably bound together, indeed they are mutually constitutive, in what he famously called by the conjoined term *pouvoir/savoir*, or power/knowledge.

References

Anderson, Benedict. 1991. *Imagined Communities: Reflections on the Origins and Spread of Nationalism.* 2d rev. ed. New York: Verso.

Bateson, Gregory. 1972. *Steps to an Ecology of Mind.* New York: Ballantine Books.

Blatter, Joachim. 1998. "Entgrenzung der Staatenwelt? Politische Institutionenbildung in Grenzüberschreitenden Regionen in Europa und Nordamerika." Ph.D. diss., Institute for Political Science, Martin-Luther-University, Halle-Wittenberg, Germany.

Castells, Manuel. 1989. *The Informational City.* Cambridge, Mass.: Blackwell.

Castells, Manuel. 1996. *The Rise of the Network Society.* Oxford: Blackwell.

Elkins, David J. 1995. *Beyond Sovereignty: Territory and Political Economy in the Twenty-First Century.* Toronto: University of Toronto Press.

Foucault, Michel. 1995. *Discipline and Punish: The Birth of the Prison*, trans. Alan Sheridan. 2d ed. New York: Vintage Books.

Geertz, Clifford. 1983. *Local Knowledge: Further Essays in Interpretive Anthropology*. New York: Basic Books

Giddens, Anthony. 1990. *The Consequences of Modernity*. Stanford, Calif.: Stanford University Press.

Habermas, Jürgen. 1987. *The Philosophical Discourse of Modernity*. Cambridge, U.K.: Polity Press.

Habermas, Jürgen. 1996. *Between Facts and Norms: Contributions to a Discourse Theory of Law and Democracy*, trans. William Rehg. Cambridge: MIT Press.

Habermas, Jürgen. 1998. "Jenseits des Nationalstaates? Bemerkungen zu Folgeproblemen der wirtschaftlichen Globalisierung." In Ulrich Beck (ed.), *Politik der Globalisierung*, pp. 67–85. Frankfurt am Main: Suhrkamp.

Heidegger, Martin. 1977. *The Question Concerning Technology and Other Essays*, trans. William Lovitt. New York: Harper and Row.

Hirst, Paul Q., and Grahame Thompson. 1996. *Globalization in Question: The International Economy and the Possibilities of Governance*. Cambridge, UK: Polity Press; Oxford, U.K., and Cambridge, Mass.: Blackwell Publishers.

Lyotard, Jean-Francois. 1985. *The Post-Modern Condition*. Minneapolis: University of Minnesota Press.

McAfee, Kathleen. forthcoming. "Selling Nature to Save It." *Environment and Society*.

Ross, Andrew. 1994. *The Chicago Gangster Theory of Life: Nature's Debt to Culture*. London, New York: Verso.

Sale, Kirkpatrick. 1998. "Liquid Asset." *Nation,* digital ed., May 11 ⟨http://www.thenation.com/⟩.

Santos, Boaventura de Sousa. 1995. *Toward a New Common Sense: Law, Science, and Politics in the Paradigmatic Transition*. New York: Routledge.

Contributors

Joachim Blatter is Assistant Professor, Department of Politics and Management, University of Konstanz, Germany. He is the author of a number of journal articles on cross-border regions.

Joseph F. DiMento is Professor of Law and Society, Planning, and Environmental Analysis at the University of California, Irvine. He is the author of numerous journal articles and several books, including *Environmental Law and American Business: Dilemmas of Compliance*.

Pamela M. Doughman is a postdoctoral fellow in the Department of Government and Politics at the University of Maryland, College Park. Among her publications is the coauthored (with Joseph F. DiMento) 1998 article "Soft Teeth in the Back of the Mouth: The NAFTA Environmental Side Agreement Implemented."

Paula Garb is Associate Adjunct Professor of Social Ecology at the University of California, Irvine. She is the author of books and articles about Abkhazia and other parts of the Caucasus based on her extensive anthropological research in the region since 1979.

María Rosa García-Acevedo is Assistant Professor in the Department of Political Science at California State University, Northridge. Her publications include articles and book chapters on U.S.-Mexico relations, environmental issues of the U.S.-Mexico border, and the Mexican/Latin American diaspora in the United States.

David McDermott Hughes is Assistant Professor of Human Ecology, Cook College, Rutgers University. He has worked in the field of conservation and development in Southern Africa for more than ten years and is the author of "Mapping the Hinterland: Land Rights, Timber, and Territorial Politics in Mozambique."

Helen Ingram is Warmington Endowed Chair in the School of Social Ecology at the University of California at Irvine. Among her nine books and more than eighty articles and book chapters, roughly two-thirds deal with water resources, including a number of studies of transboundary issues. Her coauthored (with Nancy R. Laney and David M. Gillilan) 1995 book *Divided Waters: Bridging the*

U.S.-Mexico Border and her 1988 coauthored (with Suzanne L. Fiederlein) article "Traversing Boundaries: A Public Policy Approach to the Analysis of Foreign Policy" exemplify some of the intellectual roots of this volume.

Suzanne Lorton Levesque received her Ph.D. in Environmental Health Science and Policy from the University of California, Irvine, in June 2000. Her dissertation research focused on the interactions among scientific knowledge and expertise and environmental advocacy within the network of participants involved in the Yellowstone to Yukon Conservation Initiative.

Richard Perry is Assistant Professor in the Department of Criminology, Law and Society at the University of California, Irvine. He also holds appointments in the Department of Environmental Analysis and the Interdisciplinary Program in Native American Studies. Among his publications is the forthcoming coedited book *Globalization and Governmentalities*.

Kathleen M. Sullivan is a Ph.D. candidate at the University of California, Santa Barbara. Her research focuses on political ecological relationships and the mass media and public debates about transnational environmental governance.

John M. Whiteley is Professor of Social Ecology in the Department of Environmental Analysis and Design at the University of California at Irvine. His most recent scholarship involves practical peace building in the ethnic/national disputes in the Caucasus region of the former Soviet Union. He is the coauthor (with Russell J. Dalton, Paula Garb, Nicholas P. Lovrich, and John C. Pierce) of *Critical Masses: Citizens, Nuclear Weapon Production, and Environmental Destruction in the United States and Russia.*

Index

Page numbers in italics indicate illustrations.